Nanostructure Science and Technology

Series Editor

David J. Lockwood, FRSC
National Research Council of Canada, Ottawa, Canada

Nanostructure science and technology now forms a common thread that runs through all physical and materials sciences and is emerging in industrial applications as nanotechnology. The breadth of the subject material is demonstrated by the fact that it covers and intertwines many of the traditional areas of physics, chemistry, biology, and medicine. Within each main topic in this field there can be many subfields. For example, the electrical properties of nanostructured materials is a topic that can cover electron transport in semiconductor quantum dots, self-assembled molecular nanostructures, carbon nanotubes, chemically tailored hybrid magnetic-semiconductor nanostructures, colloidal quantum dots, nanostructured superconductors, nanocrystalline electronic junctions, etc. Obviously, no one book can cope with such a diversity of subject matter. The nanostructured material system is, however, of increasing significance in our technology-dominated economy and this suggests the need for a series of books to cover recent developments.

The scope of the series is designed to cover as much of the subject matter as possible – from physics and chemistry to biology and medicine, and from basic science to applications. At present, the most significant subject areas are concentrated in basic science and mainly within physics and chemistry, but as time goes by more importance will inevitably be given to subjects in applied science and will also include biology and medicine. The series will naturally accommodate this flow of developments in the sciences and technology of nanostructures and maintain its topicality by virtue of its broad emphasis. It is important that emerging areas in the biological and medical sciences, for example, not be ignored as, despite their diversity, developments in this field are often interlinked. The series will maintain the required cohesiveness from a judicious mix of edited volumes and monographs that while covering subfields in depth will also contain more general and interdisciplinary texts.

Thus the series is planned to cover in a coherent fashion the developments in basic research from the distinct viewpoints of physics, chemistry, biology, and materials science and also the engineering technologies emerging from this research. Each volume will also reflect this flow from science to technology. As time goes by, the earlier series volumes will then serve as reference texts to subsequent volumes.

Ramachandra Naik • H P Nagaswarupa
Ganesh Koyyada • K G Basava Kumar
Editors

Biocompatible Nanomaterials for Smart Sensors

Environmental and Biomedical Applications

Springer

Editors
Ramachandra Naik
Department of Applied Sciences
New Horizon College of Engineering
Bengaluru, Karnataka, India

H P Nagaswarupa
Department of Studies in Chemistry
Davangere University
Davangere, Karnataka, India

Ganesh Koyyada
School of Chemical Engineering
Yeungnam University
Gyeongsan, Korea (Republic of)

K G Basava Kumar
ATAL-IDEA LAB, Global Campus
Visvesvaraya Technological University
Belagavi, Karnataka, India

ISSN 1571-5744 ISSN 2197-7976 (electronic)
Nanostructure Science and Technology
ISBN 978-3-032-01744-4 ISBN 978-3-032-01745-1 (eBook)
https://doi.org/10.1007/978-3-032-01745-1

This Springer imprint is published by the registered company Springer Nature Switzerland AG
The registered company address is: Gewerbestrasse 11, 6330 Cham, Switzerland

Preface

The advent of nanotechnology has revolutionized the landscape of material science, especially in the domain of sensor development for environmental and biomedical applications. In recent years, the emphasis on biocompatible nanomaterials has grown exponentially, owing to their inherent advantages such as low toxicity, enhanced surface area, tunable physicochemical properties, and environmentally sustainable synthesis routes. This book, *Biocompatible Nanomaterials for Smart Sensors: Environmental and Biomedical Applications*, is an outcome of a collaborative effort to bring together recent innovations, challenges, and future directions in this dynamic and interdisciplinary field.

The primary objective of this book is to provide a comprehensive overview of the synthesis, characterization, and practical deployment of biocompatible nanomaterials in various sensor platforms. The chapters included in this volume encompass a broad spectrum of topics—from fundamental principles to advanced applications. The contents span the development of sensors for heavy metal ion detection, biomolecule analysis, pesticide monitoring, pharmaceutical residue identification, and integration of artificial intelligence and IoT in sensing systems.

The book is structured into thirteen chapters, starting with foundational insights into nanomaterials and multifunctional sensors, followed by advanced synthesis methodologies and characterization techniques. Special focus has been given to green synthesis strategies and the practical deployment of these nanomaterials in real-world sensor applications. Chapters also address scalability, commercialization, and translational research, making the book highly relevant for researchers, industrial practitioners, academicians, and students engaged in materials science, chemistry, biomedical engineering, environmental science, and sensor technology.

This book is the result of a truly global academic collaboration. Contributors from India, Korea, Ireland, Italy, Saudi Arabia, Ethiopia, Malaysia, Sri Lanka, and Chile have shared their valuable expertise, reflecting the global significance and interdisciplinary nature of this research area.

We extend our heartfelt gratitude to all the contributing authors for their dedication and scholarly input. We also acknowledge the strong academic and institutional support received from New Horizon College of Engineering, Davangere University,

Yeungnam University, and Visvesvaraya Technological University, whose encouragement and infrastructure have significantly contributed to the successful compilation of this volume.

We are thankful to Springer for recognizing the relevance of this topic and supporting the publication of this work.

We hope this volume serves as a valuable reference for readers seeking to understand and advance the field of biocompatible nanomaterial-based smart sensors for a safer, cleaner, and healthier future.

Bengaluru, India	Ramachandra Naik
Davangere, India	H P Nagaswarupa
Gyeongsan, South Korea	Ganesh Koyyada
Belagavi, India	K G Basava Kumar

Contents

Chapter 1
Introduction to Biocompatible Nanomaterials

V. V. Deshmukh

Abstract The field of nanotechnology has revolutionized the landscape of biomedical science by introducing materials that operate at the molecular and cellular level. Among these, biocompatible nanomaterials have emerged as a pivotal class of substances, offering exceptional promise for diagnostic, therapeutic, and regenerative applications. This chapter provides a brief yet comprehensive overview of biocompatible nanomaterials, focusing on their key properties, types, synthesis, and applications in biomedical science. It traces the historical development of the field and highlights essential characteristics such as biocompatibility, biodegradability, and surface functionalization. Various classes of nanomaterials—metals, polymers, lipids, carbon-based, and natural biomaterials—are discussed along with their synthesis methods, including green approaches. Characterization techniques and safety evaluations, including in vitro and in vivo assessments, are outlined. The chapter also addresses recent advancements, regulatory aspects, and future perspectives in the safe and effective use of biocompatible nanomaterials in healthcare.

Keywords Biocompatible nanomaterials · Biomedical applications · Green synthesis

1.1 Introduction

Nanotechnology, a rapidly evolving field, deals with materials and structures on the nanometer scale—typically ranging from 1 to 100 nm. As technology has advanced, our understanding and manipulation of materials at this scale have significantly improved, giving rise to numerous applications in electronics, energy, and especially medicine. One such critical branch is the development and use of biocompatible nanomaterials [1].

V. V. Deshmukh (✉)
Department of Physics, Shri Shivaji Science College, Amravati, Maharashtra, India

R. Naik et al. (eds.), *Biocompatible Nanomaterials for Smart Sensors*, Nanostructure Science and Technology,
https://doi.org/10.1007/978-3-032-01745-1_1

Biocompatible nanomaterials refer to nanoscale substances that can interact with biological systems without eliciting harmful immune responses or toxic effects. These materials are designed to mimic or integrate with biological tissues and are extensively used in drug delivery, tissue engineering, diagnostics, and biosensing [2].

Biocompatibility is the ability of a material to perform its desired function within a biological system without causing adverse reactions. When materials are engineered at the nanoscale, their surface area increases dramatically, making them more reactive and enhancing their interaction with cells and tissues [3].

Figure 1.1 emphasizes that biocompatibility is a multifactorial concept governed by both material characteristics and biological context, crucial for designing safe and effective nanomaterials for medical applications. When nanoparticles enter the human body, they are identified by the immune system as foreign entities, much like bacteria, viruses, or other unfamiliar substances. This recognition triggers a cascade of biological responses designed to neutralize or eliminate the intruding material. These responses may include activation of immune cells such as macrophages, the release of inflammatory mediators, and engagement of clearance mechanisms through the liver, kidneys, or lymphatic system. While these defense strategies are essential for protecting the body against harmful invaders, they can also pose significant challenges in the context of nanomedicine [4].

If the interaction between nanoparticles and the biological environment leads to exaggerated immune responses, prolonged inflammation, or damage to healthy tissues, the nanoparticles are considered bio-incompatible. Such adverse reactions can result in toxicity, unintended immune activation, oxidative stress, or even long-term complications such as thrombosis (blood clot formation) or tumorigenesis (development of cancer) [5].

Fig. 1.1 Schematic representation of biocompatibility

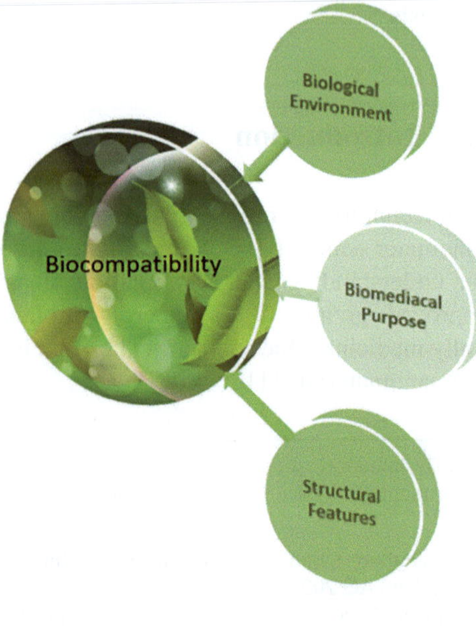

For a nanomaterial to be deemed highly biocompatible, it must be able to function within the body for its intended purpose without causing undesirable side effects. This includes avoiding cytotoxicity (harm to cells), immunogenicity (triggering immune responses), and genotoxicity (damage to genetic material). In addition, the material should not interfere with normal physiological functions, nor should it accumulate in vital organs in a way that could impair their activity over time [6].

The biological fate and compatibility of nanoparticles are heavily influenced by several factors, including their size, shape, surface charge, chemical composition, and surface modifications. For instance, nanoparticles with hydrophilic (water-attracting) surfaces or those coated with biocompatible polymers like polyethylene glycol (PEG) tend to evade rapid clearance and reduce protein adsorption, leading to longer circulation times in the bloodstream. On the other hand, uncoated or hydrophobic nanoparticles may rapidly bind with plasma proteins, forming a "protein corona" that alters their identity and influences how they are perceived by the immune system [7, 8].

Moreover, the interaction of nanoparticles with cellular membranes, lipid bilayers, and various intracellular compartments can affect their distribution, uptake, and degradation. Their ability to cross biological barriers, such as the blood-brain barrier or intestinal lining, is particularly important for therapeutic and diagnostic applications. Therefore, understanding and carefully engineering the surface chemistry and physical characteristics of nanoparticles is critical to achieving optimal biocompatibility [9, 10].

Therefore, the compatibility of nanoparticles with the biological environment is not just a matter of material selection but involves a comprehensive understanding of how these materials behave in complex biological systems. Only through such thoughtful design and evaluation can we ensure that nanoparticles perform their intended functions safely and effectively, opening up new possibilities in biomedical science, diagnostics, and therapeutics.

1.2 Historical Background and Evolution

The concept of biocompatibility has deep roots in the history of medicine, tracing back to ancient times when materials like wood, ivory, and metals were used to replace or support damaged body parts. However, the systematic exploration of materials compatible with biological systems began in earnest only in the twentieth century, driven by the needs of modern surgery and prosthetics.

In the early 1900s, materials such as stainless steel, gold, and silicone were introduced into medical applications. These were among the first materials selected for their relative inertness and ability to coexist with human tissue without causing severe rejection or toxic effects. Their use in dental fillings, bone plates, and heart valves marked the beginning of a more scientific approach to medical materials [11].

The term biocompatibility itself gained prominence in the 1970s, as researchers began to focus on the interaction between synthetic materials and living tissues. It became increasingly clear that for a material to succeed in a biomedical setting, it must do more than perform a mechanical function—it must also interact harmoniously with the body at the cellular and molecular levels.

The rise of nanotechnology in the late twentieth century introduced a paradigm shift in material science. With the ability to manipulate matter at the atomic and molecular scale, researchers began exploring how nanoscale materials could interact with biological systems in entirely new ways. The unique properties of nanomaterials—such as their high surface-area-to-volume ratio, tunable surface chemistry, and quantum effects—opened the door to applications that were not possible with bulk materials [12–14].

In the 1990s, the integration of nanotechnology into biomedical research gave birth to what is now known as nanomedicine. Scientists began designing nanoparticles that could deliver drugs directly to disease sites, serve as contrast agents in imaging, or act as carriers for genes and proteins. These advances demanded materials that were not only functional but also safe and biocompatible [15, 16].

As research progressed, a new generation of nanomaterials emerged—engineered specifically to minimize toxicity and maximize compatibility with human tissues. Natural polymers such as chitosan, alginate, and collagen were incorporated into nanoparticle formulations due to their inherent biocompatibility. Simultaneously, synthetic polymers like polylactic acid (PLA), polyglycolic acid (PGA), and their copolymers (PLGA) were developed for use in biodegradable nanocarriers [17]. Surface modification techniques became more sophisticated, allowing scientists to "camouflage" nanoparticles from the immune system or direct them to specific cell types. This led to the creation of smart nanomaterials, capable of responding to biological cues like pH, temperature, or enzymatic activity [18].

In recent years, the focus has shifted from simply avoiding toxicity to actively promoting positive biological interactions. Researchers now strive to design bioactive nanomaterials that not only do no harm but also support processes such as tissue regeneration, immune modulation, and cellular signaling. These materials are at the heart of innovations in tissue engineering, precision medicine, and minimally invasive therapies [19, 20].

Moreover, the rise of personalized medicine has encouraged the development of customizable nanomaterials tailored to individual patient profiles, minimizing adverse reactions and improving treatment outcomes. Regulatory bodies have also begun to establish more rigorous frameworks to evaluate the safety and efficacy of nanomaterials in clinical settings.

1.3 Key Properties of Biocompatible Nanomaterials

Biocompatible nanomaterials are designed to work safely inside the human body. For these materials to be useful in medical applications, they must have certain important properties. Some of the most essential ones include biocompatibility, biodegradability, and surface functionalization.

Biocompatibility is the most important property of any material used in the body. It means that the material should not cause harm to living tissues. When a nanomaterial is placed inside the body, it should not cause any negative reaction like inflammation, allergy, infection, or damage to cells. A truly biocompatible nanomaterial is one that the body can accept without triggering harmful immune responses. For example, if a nanoparticle is used to carry a drug to a tumor, it should reach the tumor safely without harming healthy cells. It should also not cause blood clots or interfere with normal body functions. Scientists carefully test nanomaterials to ensure they are safe and well-tolerated before using them in medicines or implants [21].

Biodegradability means the ability of the material to break down naturally in the body over time. After the nanomaterial has done its job—such as delivering a drug—it should gradually dissolve or be broken down by the body's natural systems. This helps avoid the need for surgical removal or long-term side effects. Biodegradable nanomaterials are especially useful because they reduce the risk of material buildup in organs like the liver or kidneys. They are broken down into small, harmless parts that can be removed through urine or other natural processes. Materials like polylactic acid (PLA) and chitosan are examples of biodegradable materials commonly used in biomedical applications [22, 23].

Surface functionalization is the process of modifying the outer surface of a nanomaterial. This helps the nanomaterial do specific tasks inside the body. For example, scientists can attach targeting molecules to the surface so the nanoparticle can find and stick to a certain type of cell, like a cancer cell. The surface of a nanomaterial can also be changed to make it more stable in blood, reduce immune system reactions, or improve how long it stays in the body. This is done by coating the particle with special materials like polyethylene glycol (PEG), which helps the particle move freely in the bloodstream without being attacked by immune cells. Functionalization also helps the nanomaterial communicate better with cells, tissues, and biological systems. This increases the effectiveness of drug delivery, imaging, and other medical uses [24].

1.4 Types of Biocompatible Nanomaterials

Figure 1.2 shows the classification of biocompatible materials into five major types: metal and metal-based, carbon-based, polymeric, lipid-based, and natural materials.

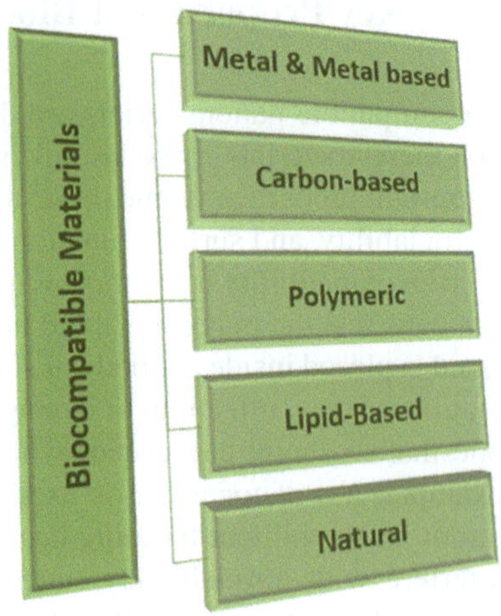

Fig. 1.2 Types of biocompatible materials

Metal-based nanomaterials are a prominent class of engineered nanostructures made from metallic elements or metal oxides. Their unique physicochemical properties, such as large surface area, electrical conductivity, optical tunability, and magnetic behavior, make them highly valuable in various biomedical applications, including diagnostics, therapeutics, imaging, and antimicrobial coatings. However, ensuring biocompatibility through surface functionalization or coating is essential to avoid cytotoxicity and ensure safe use in biological environments.

1. Gold nanoparticles: Inert and non-toxic; easily functionalized with biomolecules like DNA, antibodies, or peptides. Used in cancer photothermal therapy, drug delivery, and biosensing due to their surface plasmon resonance (SPR) properties. Regarding the application of gold NPs in biomedicine (sensing and drug delivery applications), they should be easily dispersible at neutral pH and should be stable[25, 26].
2. Silver nanoparticles: Antimicrobial activity is useful in wound dressings and implants, but requires careful dose management due to possible cytotoxicity. Used in antibacterial coatings, dental applications, and topical formulations [27].
3. Iron oxide nanoparticles: Iron oxide NPs must be highly crystalline, monodisperse, and water soluble, providing high magnetization values, reproducible quality, and good biocompatibility under biological conditions [28, 29]. Used in magnetic resonance imaging (MRI), magnetic hyperthermia, and targeted drug delivery through external magnetic fields. FDA-approved as MRI contrast agents (e.g., Ferumoxytol); biodegradable into ferritin.
4. Zinc oxide (ZnO) and titanium dioxide (TiO_2) Nanoparticles: ZnO is biodegradable and exhibits selective cytotoxicity toward cancer cells. TiO_2 is widely used in sunscreens and has low toxicity. Both used in antibacterial agents, photocatalytic cancer therapy, UV-protection in cosmetics, and bone implants [30].
5. Bimetallic and alloy nanoparticles: Combine the properties of two or more metals for enhanced functionality and reduced toxicity. These combinations can be

in the form of core–shell structures, heterostructures, or homogenous alloys. The synergistic effects between the constituent metals often lead to enhanced physicochemical, optical, magnetic, and catalytic properties compared to their monometallic counterparts. Ag–Cu and Ag–Zn, display superior antibacterial activity, Au–Pd and Au–Pt nanoparticles have been used in photothermal therapy (PTT) and photodynamic therapy (PDT). Pt–Fe and Au–Fe bimetallic nanoparticles can be magnetically guided to tumor sites and subsequently heated for hyperthermia therapy, while also carrying drug payloads. Au–Ag nanoparticles are widely used in surface-enhanced Raman scattering (SERS) for sensitive biomarker detection, and Fe–Au particles serve dual purposes in MRI contrast and photothermal therapy [31–33].

Carbon-based nanomaterials are a diverse class of nanostructures composed primarily of carbon atoms arranged in various allotropic forms. Due to their exceptional mechanical strength, electrical conductivity, high surface area, and tunable surface chemistry, they have found extensive applications in biomedicine, diagnostics, biosensing, tissue engineering, and drug delivery. Their biocompatibility can be significantly enhanced through surface functionalization, which allows for safe interaction with biological environments.

1. Carbon nanotubes (CNTs): Cylindrical structures composed of rolled graphene sheets (single-walled or multi-walled). Used in drug delivery, neural interface development, and biosensors. Raw CNTs are hydrophobic and may be cytotoxic; however, functionalization with polymers, PEG, or biomolecules significantly improves dispersion and reduces toxicity [34].
2. Graphene and graphene oxide (GO): Single layers of carbon atoms in a hexagonal lattice. GO contains oxygen functional groups, enhancing hydrophilicity. Used in anticancer drug delivery, scaffolds for tissue regeneration, and photothermal therapy. GO is more biocompatible than pristine graphene. Functionalized GO can interact safely with cells and tissues [35, 36]. However, the effects of graphene nanomaterials on cell viability depend on various factors and experimental conditions.
3. Carbon dots (CDs) and graphene quantum dots (GQDs): Carbon dots are quasi-spherical nanoparticles composed of amorphous or crystalline carbon. They exhibit strong excitation-dependent fluorescence, which is highly tunable by modifying their surface chemistry or particle size. CDs are excellent fluorescent probes for imaging cells and tissues due to their bright emission and low toxicity. CDs synthesized from citric acid have been used to label HeLa cells with high resolution and no cytotoxic effects. Functionalized CDs can carry drugs like doxorubicin and release them in response to pH or enzymatic changes in cancer cells. GQDs are small fragments of graphene with lateral sizes <10 nm. They retain a sp^2-hybridized carbon lattice and possess quantum confinement and edge effects, which are responsible for their photoluminescence. GQDs can generate reactive oxygen species under light, useful for killing cancer cells. GQDs functionalized with polyethylene glycol (PEG) and folic acid target cancer cells and enable effective PDT. GQDs functionalized with metal ions or biomolecules

can detect glucose, dopamine, or even environmental toxins at nanomolar concentrations [37–40].

Polymeric nanomaterials are a versatile class of biocompatible nanostructures widely employed in biomedical applications, especially in drug delivery, tissue engineering, gene therapy, and diagnostics. These nanomaterials are either natural or synthetic polymers, and their biodegradability, non-toxicity, and tunable physicochemical properties make them highly suitable for interactions with biological systems. They are classified into main two groups natural and synthetic.

1. Natural polymers are derived from biological sources and are inherently biocompatible. Examples include:
   (a) Chitosan: A polysaccharide derived from chitin, widely used in drug delivery systems due to its biodegradability, mucoadhesiveness, and ability to form polyelectrolyte complexes. It is especially effective for oral and nasal delivery of proteins and genes [41].
   (b) Alginate: Extracted from brown seaweed, alginate forms hydrogels in the presence of calcium ions and is used in encapsulating drugs, probiotics, and cells in tissue scaffolds [42].
   (c) Gelatin and collagen: Derived from animal tissues, these proteins support cell attachment and proliferation, making them suitable for tissue regeneration and wound healing applications.
2. Synthetic polymers are engineered for specific applications and offer greater control over degradation rates, molecular weight, and surface properties:
   (a) PLGA (poly(lactic-co-glycolic acid)): One of the most commonly used synthetic polymers in nanomedicine. It is approved by the FDA due to its safe degradation into lactic and glycolic acid. PLGA nanoparticles are ideal for sustained and controlled release of chemotherapeutic drugs [43].
   (b) PEG (polyethylene glycol): Frequently used as a surface modifier ("PEGylation") to enhance nanoparticle stability, solubility, and circulation time. It prevents opsonization and prolongs systemic half-life [44].
   (c) PCL (polycaprolactone): A slowly degrading polymer used in long-term drug delivery and biomedical implants. Its mechanical strength also supports use in 3D scaffolds for bone tissue engineering.
3. Biomedical applications.
   (a) Targeted drug delivery: Polymeric nanocarriers are used to direct drugs to specific tissues or tumors, reducing side effects and enhancing efficacy.
   (b) Gene therapy: Polymers like chitosan or PEI can deliver DNA or RNA efficiently, avoiding the risks associated with viral vectors.
   (c) Tissue engineering: Biodegradable scaffolds made from polymers like PLGA and gelatin promote cell attachment and tissue growth.
   (d) Imaging and diagnostics: Functionalized polymeric nanoparticles are being developed as contrast agents for MRI, CT, and fluorescence imaging.

Lipid-based biocompatible nanomaterials are a versatile class of nanoscale drug delivery systems characterized by their excellent biocompatibility, biodegradability, and ability to encapsulate a wide range of therapeutic agents. These nanocarriers have gained significant attention in biomedical applications, including cancer therapy, gene delivery, and vaccine development [45–47]. Types are:

- Liposomes: Spherical vesicles composed of one or more phospholipid bilayers, capable of encapsulating both hydrophilic and hydrophobic drugs. Their structural similarity to biological membranes facilitates cellular uptake and reduces immunogenicity.
- Solid lipid nanoparticles (SLNs): Comprised of solid lipids stabilized by surfactants, SLNs offer improved drug stability and controlled release profiles. They are particularly useful for delivering lipophilic drugs.
- Nanostructured lipid carriers (NLCs): These are second-generation lipid nanoparticles that incorporate both solid and liquid lipids, enhancing drug loading capacity and preventing drug expulsion during storage.
- Lipid nanoemulsions: Oil-in-water emulsions stabilized by surfactants, suitable for intravenous administration of poorly water-soluble drugs.

Despite their promising attributes, challenges persist in the clinical translation of lipid-based nanocarriers. Issues related to large-scale manufacturing, stability during storage, and potential immunogenicity necessitate ongoing research and optimization. Furthermore, regulatory considerations and the need for comprehensive characterization underscore the importance of standardized protocols in the development of these nanomaterials.

Natural Biocompatible Nanomaterials: Natural biocompatible nanomaterials are nanoscale materials derived from natural sources that are inherently compatible with biological systems. These materials are biodegradable, non-toxic, and often possess intrinsic biological activities, making them ideal for various biomedical applications such as drug delivery, tissue engineering, and biosensing [48, 49].

Figure 1.3 shows various natural biocompatible nanomaterials derived from biological sources. One prominent example is chitosan, a polysaccharide obtained from the deacetylation of chitin found in crustacean shells. Chitosan exhibits excellent biocompatibility, biodegradability, and antimicrobial properties. It has been widely used to develop nanoparticles for drug delivery systems, where it enhances drug stability and facilitates controlled release. Additionally, chitosan-based scaffolds have been employed in tissue engineering due to their ability to support cell adhesion and proliferation [50–52].

Another significant natural nanomaterial is alginate, extracted from brown seaweed. Alginate forms hydrogels in the presence of divalent cations like calcium, creating a three-dimensional network suitable for encapsulating cells and bioactive molecules. This property makes alginate-based hydrogels valuable in wound healing and tissue regeneration applications [53].

Silk fibroin, derived from the silk of the *Bombyx mori* silkworm, is a protein-based nanomaterial known for its remarkable mechanical strength and biocompatibility. Silk fibroin nanoparticles have been explored for drug delivery, offering

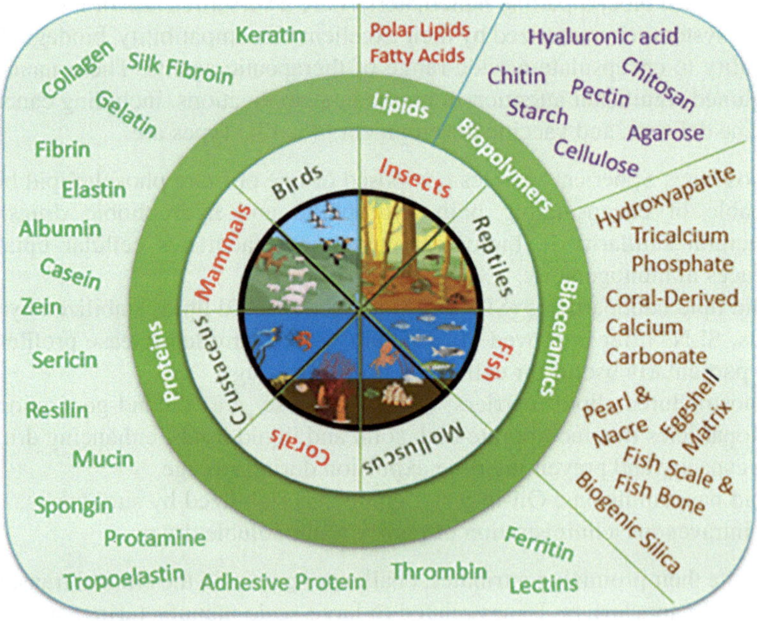

Fig. 1.3 Natural biocompatible material

controlled release profiles and minimal immunogenicity. Moreover, silk fibroin scaffolds have been utilized in bone and cartilage tissue engineering [54].

Cellulose nanocrystals (CNCs), obtained from plant cellulose, are rod-like nanoparticles with high tensile strength and surface area. CNCs are biodegradable and have been incorporated into composite materials for biomedical applications, including as reinforcing agents in hydrogels for tissue engineering [55].

Hyaluronic acid (HA), a naturally occurring glycosaminoglycan found in connective tissues, plays a crucial role in cell signaling and wound healing. HA-based nanoparticles have been developed for targeted drug delivery, particularly in cancer therapy, due to HA's affinity for CD44 receptors overexpressed in certain tumor cells.

Gelatin, a denatured form of collagen, is another natural biopolymer used to fabricate nanoparticles for drug delivery and tissue engineering. Gelatin nanoparticles are biodegradable and can be crosslinked to modulate their mechanical properties and degradation rates, making them suitable for controlled drug release applications [56].

Starch-based nanoparticles, derived from plant starch, offer advantages such as biodegradability, biocompatibility, and ease of modification. These nanoparticles have been investigated for delivering hydrophilic drugs and genes, as well as for developing edible coatings in food packaging.

Natural lipid-based nanomaterials, such as liposomes and solid lipid nanoparticles, are composed of biocompatible lipids that can encapsulate both hydrophilic and hydrophobic drugs. These systems enhance drug solubility, stability, and bioavailability, and have been employed in various therapeutic areas, including oncology and infectious diseases. Hence, natural biocompatible nanomaterials offer a versatile and sustainable platform for biomedical applications. Their inherent compatibility with biological systems, coupled with their functional properties, make them promising candidates for advancing healthcare technologies.

Table 1.1 shows the relationship between various biocompatible nanomaterials, their synthesis methods, and respective biomedical applications. It highlights how the choice of synthesis technique influences the functional properties and suitability of nanomaterials for targeted uses such as drug delivery, tissue engineering, and biosensing.

Table 1.1 Summary of literature on biocompatible nanomaterials, their synthesis methods, and biomedical applications

S. No.	Nanomaterial	Synthesis method	Application	References
1.	Bismuth ferrite ($BiFeO_3$)	Hydrothermal synthesis	Biomedical imaging and diagnostics	[57]
2.	Titanium nanosheets	Liquid-phase exfoliation	Photothermal cancer therapy	[58]
3.	Copper-zinc metal-organic frameworks (MOFs)	Electrospinning and coating	Bone tissue engineering	[59]
4.	Cu^{2+}-functionalized reduced graphene oxide	Functionalization with Cu^{2+}	Antibacterial applications	[60]
5.	Silver-loaded mesoporous silica nanoparticles	Impregnation method	Antimycobacterial agents against *M. tuberculosis*	[61]
6.	Curcumin-Cu and curcumin-Ag nanoparticles	Niosome encapsulation	Antibacterial and anti-biofilm activities	[62]
7.	Nickel (II) complexes	Intercalation	Antioxidant and anticholinergic activity	[63]
8.	GO-cellulose acetate scaffold	Electrospinning	Tissue engineering	[64]
9.	Functionalized MWCNTs	Chemical functionalization	Drug delivery systems	[65]
10.	rGO-coated collagen scaffolds	Chemical reduction	Bone tissue engineering	[66]
11.	Graphene quantum dot formulation	Synthesized from glucosamine	Cancer imaging and redox-based drug delivery	[67]
12.	BMP-2 loaded PLGA nanoparticles	Double-emulsion (W/O/W) solvent evaporation technique	Bone regeneration	[68]
13.	Agarose-chitosan hybrid materials with Ag_2O nanoparticles	Dissolution in ionic liquid, regeneration, and sol-gel transition	Antimicrobial activity	[69]

(continued)

Table 1.1 (continued)

S. No.	Nanomaterial	Synthesis method	Application	References
14.	Solid lipid nanoparticles	Phase inversion temperature (PIT) method	Topical delivery of hydrogen peroxide-detecting chemiluminescent dyes	[70]
15.	Lipid–polymer hybrid nanoparticles	Single-step sonication method	Drug delivery applications	[71]
16.	Lipid nanoparticles	High-pressure homogenization (HPH)	Drug delivery applications	[72]
17.	Lipid nanoparticles	Microfluidic synthesis	Drug delivery	[73]
18.	Chitosan nanoparticles	Ionic gelation	Drug delivery and antimicrobial	[74]
19.	Alginate nanoparticles	Solvent evaporation	Drug delivery and wound healing	[75]
20.	Silk fibroin nanoparticles	Electrospinning	Tissue engineering and drug delivery	[76]

1.5 Synthesis Techniques for Biocompatible Nanomaterials

The synthesis of biocompatible nanomaterials has become a pivotal aspect of research due to their diverse applications in medicine, biotechnology, and environmental remediation. These materials must adhere to strict biological safety standards while demonstrating high efficiency in various applications. Nanomaterials can be synthesized using different approaches, each offering unique advantages based on the desired properties. Figure 1.4 illustrates the major synthesis techniques for biocompatible nanomaterials, categorized into top-down, bottom-up, and green synthesis methods.

1.5.1 Physical Methods

These methods primarily rely on mechanical, thermal, or other non-chemical techniques to shape and manipulate nanomaterials.

- High-energy ball milling: This technique is like grinding a pile of rocks into fine powder. It uses intense mechanical forces to break down bulk materials into nanometer-sized particles. For example, the synthesis of metallic nanomaterials, like iron or gold nanoparticles, can be achieved through repeated collisions between balls in a rotating container. This method is particularly useful for creating powders that can be used in biomedical implants [77].

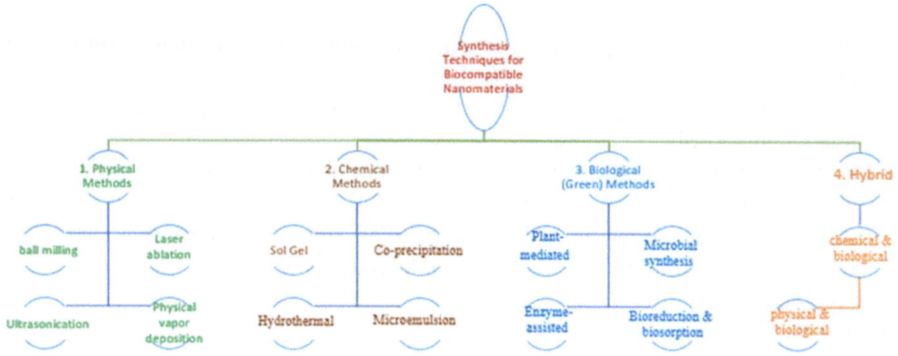

Fig. 1.4 Synthesis techniques for biocompatible nanomaterials

- Laser ablation: Imagine using a powerful laser beam to vaporize a solid material, forming a cloud of nanoparticles. This technique involves focusing high-powered lasers on the surface of materials, causing them to evaporate and cool, forming fine nanoparticles. For instance, silver nanoparticles can be generated using this method, which then exhibit excellent antimicrobial properties, making them useful in wound healing [78].
- Ultrasonication: Think of this method as using high-frequency sound waves to break apart substances into tiny particles. By immersing materials in a solvent and applying intense ultrasound, large clusters can be shattered into nanometer-sized pieces. This technique is commonly employed in the dispersion of carbon nanotubes, which are then used in drug delivery systems due to their biocompatibility and ability to interact with biological tissues [79].
- Physical vapor deposition (PVD): This method involves the vaporization of a solid material into a gas, which then condenses onto a substrate to form thin films or nanoparticles. PVD is widely used to produce biocompatible coatings, such as titanium or ceramic films for use in implants, offering enhanced durability and corrosion resistance [80].

1.5.2 *Chemical Methods*

Chemical methods often involve the use of various reactants and solvents to control the properties of nanomaterials at the molecular level.

- Sol-gel synthesis: In this approach, a liquid "sol" transitions into a gel-like solid through hydrolysis and condensation reactions. A common example is the synthesis of silica nanoparticles, which can be used for drug delivery systems. The sol-gel process provides control over the nanomaterial's size, shape, and surface properties, making it a versatile method for fabricating biocompatible materials [81].

- Co-precipitation: This technique is akin to making a "chemical stew" where two or more materials are precipitated out of a solution to form nanoparticles. For example, when iron salts are mixed with a base, iron oxide nanoparticles are formed, which have applications in magnetic resonance imaging (MRI) as contrast agents [82].
- Hydrothermal synthesis: A process where nanomaterials are synthesized under high pressure and temperature in aqueous solutions. The hydrothermal method is often used to create metal oxide nanoparticles, such as titanium dioxide (TiO_2), which can be employed as biocompatible coatings in dental implants or bone prosthetics [83].
- Microemulsion techniques: Imagine mixing oil, water, and surfactants to create tiny droplets. These droplets form microemulsions, where the size and shape of the droplets can be controlled to form nanoparticles. This approach is ideal for synthesizing materials like gold or silver nanoparticles, which are used in cancer therapy due to their ability to target specific cells [84].
- Chemical vapor deposition (CVD): A method in which gaseous precursors react and deposit on a substrate to form thin films or nanoparticles. A typical example is the production of carbon nanotubes, which are widely used for drug delivery and biosensing due to their high surface area and biocompatibility [85].

1.5.3 Biological (Green) Methods

As the world shifts toward sustainability, biological or green methods have gained prominence for their eco-friendly and cost-effective approach to nanomaterial synthesis.

- Plant-mediated synthesis: This process involves using plant extracts to reduce metal salts into nanoparticles. For example, silver nanoparticles can be synthesized using extracts from plants like *Azadirachta indica* (Neem), which are then employed for their antimicrobial properties in wound care [86].
- Microbial synthesis (bacteria, fungi, algae): Nature is a master at producing nanomaterials! Certain microbes, such as bacteria, fungi, and algae, can reduce metal ions to form nanoparticles. The bacteria *Bacillus subtilis* can be used to synthesize gold nanoparticles, which are highly biocompatible and can be used in medical diagnostics [87].
- Enzyme-assisted synthesis: Enzymes, the catalysts of the biological world, are harnessed to facilitate the synthesis of nanoparticles by reducing metal salts into nanoparticles at ambient conditions. For example, enzymes like phytase are used to produce nanoparticles with enhanced biocompatibility, making them ideal for drug delivery applications [88].

Bioreduction and Biosorption: This method leverages the ability of organisms like fungi and bacteria to reduce metal ions to their nanoparticulate form or to absorb them onto their surfaces. The use of fungal species in the bioreduction of gold to gold nanoparticles is a popular example, making it suitable for applications in both the medical and environmental sectors.

1.5.4 Hybrid Approaches

In the evolving world of nanotechnology, the combination of different techniques is becoming increasingly popular to enhance the properties of biocompatible nanomaterials.

- Combination of chemical and biological methods: In this hybrid approach, chemical and biological methods are integrated to optimize the synthesis of nanomaterials. For instance, plant extracts can be used to reduce metal ions in a controlled chemical environment, producing highly stable nanoparticles that benefit from both chemical precision and biological sustainability [89].
- Integration of physical and biological techniques: This hybrid method combines the robustness of physical techniques, such as laser ablation, with the eco-friendliness of biological methods like microbial synthesis. The result is the formation of nanoparticles with superior properties, such as enhanced biocompatibility and functional versatility. For example, combining laser ablation with microbial methods can yield nanoparticles with tailored surface coatings, which is beneficial for controlled drug delivery [90].

Overall, the synthesis of biocompatible nanomaterials offers a diverse array of methods, each with distinct advantages depending on the material's intended use. From the precision of physical methods to the eco-friendliness of biological approaches, the field is rapidly advancing, opening up exciting opportunities in medicine, environmental science, and beyond. The hybrid approaches further exemplify the potential for innovation, blending the best of both worlds to produce next-generation nanomaterials that meet the stringent requirements of biocompatibility and functionality.

1.6 Characterization Methods for Biocompatible Nanomaterials

The journey from synthesizing biocompatible nanomaterials to applying them in biomedical or environmental fields hinges on thorough characterization. These materials must not only possess desirable physical attributes but must also interact safely and effectively within biological systems. To evaluate their potential, researchers use a triad of characterization methods—each unveiling a different layer of their story: structure and shape, surface identity, and biological behavior.

1.6.1 Structural and Morphological Analysis

Think of this as reading the architectural blueprint of a nanomaterial. We want to know: *How big is it? What shape does it take? Does it form clusters or remain solitary?*

- Transmission electron microscopy (TEM) and scanning electron microscopy (SEM) are the microscopes of choice, zooming in down to the atomic level. For instance, gold nanoparticles synthesized via plant extracts may appear spherical with a uniform diameter of ~20 nm under TEM. This uniformity plays a role in how consistently they interact with biological cells.
- X-ray diffraction (XRD) helps identify crystalline phases and interatomic spacing. Titanium dioxide (TiO_2) nanoparticles, for example, may show clear anatase or rutile phases through XRD, revealing how the atomic lattice is arranged—a feature that affects photocatalytic activity and compatibility in bone implants.
- Atomic force microscopy (AFM), much like a blind person reading Braille, scans surfaces at the nanoscale and measures texture and height profiles. For nanocomposite coatings on implants, AFM reveals whether the surface will encourage or deter cell adhesion.

1.6.2 Surface Chemistry and Composition

Nanomaterials may be tiny, but their surfaces are chemically alive. Understanding what coats or decorates the surface is crucial to knowing how they will behave in biological environments.

- Fourier-transform infrared spectroscopy (FTIR) detects the presence of chemical groups like –OH, $-NH_2$, or –COOH. When green-synthesized silver nanoparticles are coated with plant polyphenols, FTIR can confirm their presence, showing strong peaks around 3300 cm^{-1} (O-H stretch) or 1640 cm^{-1} (C=O stretch), ensuring stability and reduced toxicity.
- X-ray photoelectron spectroscopy (XPS) dives deeper, examining elemental composition and oxidation states. A ZnO nanoparticle surface might reveal Zn^{2+} and O^{2-} peaks, confirming purity and predicting reactivity.
- Energy dispersive X-ray spectroscopy (EDS/EDX) is often paired with SEM to map where elements like Ag, Au, or Ti reside on a nanomaterial. This method is particularly helpful in verifying whether the dopants are uniformly distributed, which influences performance in biosensors.

1.6.3 Biocompatibility Testing

You can make the most beautiful nanostructure—but if it harms living tissues; It is a non-starter. Biocompatibility ensures that materials not only *survive* but also *coexist harmoniously* with biological environments.

- In vitro assays, such as MTT, LDH, or Alamar Blue, are performed on cell cultures like human fibroblasts or stem cells. These tests measure mitochondrial activity, membrane integrity, or metabolic activity. A hydrothermally synthesized cerium oxide nanoparticle may show >90% cell viability at low concentrations, indicating non-toxicity [91].
- Hemolysis tests examine if nanoparticles rupture red blood cells. Biocompatible iron oxide nanoparticles used for MRI contrast should ideally have a hemolysis ratio below 5%.
- Reactive oxygen species (ROS) assays assess oxidative stress. Materials like graphene oxide must be scrutinized for ROS generation, which can lead to inflammation or DNA damage if not controlled.
- In vivo studies, typically in animal models, explore systemic responses like inflammation, accumulation, or organ toxicity. For example, gold nanoparticles injected into mice must be tracked for clearance via the liver or kidneys and checked for inflammatory markers [92].

1.7 Toxicological and Environmental Considerations of Biocompatible Nanomaterials

As nanomaterials revolutionize medicine, energy, and environmental technologies, their tiny size belies a complex reality: what seems safe in design may pose unseen risks in biology and ecosystems. Thus, assessing their toxicological impact and environmental footprint is as essential as the materials themselves. Two pillars uphold this evaluation: biological assessments (in vitro and in vivo) and regulatory guidelines, ensuring safety for both humans and the planet [93].

1.7.1 In Vitro and In Vivo Assessments

In vitro (Latin for "in glass") assessments provide a controlled and rapid method to study how nanomaterials interact with isolated cells. In vivo tests, by contrast, evaluate these materials within living organisms to monitor systemic responses [94].

In vitro tests include:

- Cytotoxicity assays (e.g., MTT, trypan blue): Used to determine if nanoparticles inhibit cell growth or induce death. For example, silver nanoparticles synthesized from *Azadirachta indica* (neem) show dose-dependent cytotoxicity in lung epithelial cells.

- Oxidative stress assays: Detect ROS (reactive oxygen species) generation, which can trigger inflammation or apoptosis. Titanium dioxide nanoparticles often raise ROS levels under UV exposure, even if considered "inert" in dark conditions.
- Genotoxicity tests (e.g., comet assay): Assess DNA damage. Graphene oxide nanosheets, while promising for drug delivery, have shown potential to fragment DNA in human lymphocytes under certain concentrations.

In vivo models provide a real-world biological context:

- Rodent models assess long-term organ accumulation, metabolism, and clearance. Gold nanoparticles, when injected intravenously in mice, predominantly accumulate in the liver and spleen, with clearance taking weeks.
- Zebrafish embryos are increasingly used due to their transparent bodies and rapid development. They can reveal developmental toxicity or neurobehavioral changes induced by exposure to metal oxide nanoparticles.
- Aquatic organisms such as *Daphnia magna* (water fleas) are used to study environmental toxicity—useful in understanding how nanomaterials might disrupt ecosystems when they leach into water bodies.

1.7.2 Regulatory Guidelines and Safety Evaluation

Despite their novel applications, nanomaterials often float in a gray zone of regulation. The challenge lies in their variability—changing size, surface coatings, and functionality can drastically shift their behavior. Regulatory frameworks are now evolving to match the pace of innovation.

- OECD guidelines (Organisation for Economic Co-operation and Development) offer standardized tests for nanoparticle toxicity, ecotoxicity, and mutagenicity. These include protocols for inhalation studies, skin irritation, and aquatic bioaccumulation.
- REACH regulation (registration, evaluation, authorisation, and restriction of chemicals) in the European Union mandates detailed disclosure of nanomaterials used in products above 1 tonne/year, including safety data sheets and exposure models.
- FDA (US Food and Drug Administration) requires that nanomaterial-based drugs or medical devices undergo rigorous testing, including biocompatibility, sterility, and degradation products before approval.
- ISO Standards (e.g., ISO/TR 10993 series) define biocompatibility tests for medical-grade nanomaterials, ensuring that implants, coatings, or scaffolds do not cause undue harm.

- Environmental Protection Agency (EPA) has begun monitoring nanomaterials in water and air, with frameworks like the TSCA (Toxic Substances Control Act) nanomaterial inventory.

Importantly, these frameworks emphasize a "safe-by-design" approach—encouraging researchers to consider safety profiles from the very beginning of material synthesis. For instance, capping silver nanoparticles with biopolymers such as chitosan not only improves stability but also reduces cytotoxicity. Hence, toxicological and environmental assessments are not barriers but bridges—ensuring that the marvels of nanotechnology are built on safety, sustainability, and societal trust. From petri dish to planet, understanding how nanomaterials behave in diverse biological and ecological systems enables responsible innovation [95, 96].

1.8 Recent Advances and Trends in Biocompatible Nanomaterials

In the swiftly evolving realm of nanotechnology, the line between biology and engineering is becoming increasingly blurred. Today's biocompatible nanomaterials are not merely passive carriers or structural supports—they are smart, multifunctional entities capable of healing, sensing, and adapting to complex biological environments. Recent advances are transforming these materials from simple bio-inert coatings into dynamic systems that participate actively in diagnostics, therapy, and regenerative processes.

1.8.1 Smart and Stimuli-Responsive Nanomaterials

One of the most groundbreaking shifts is the emergence of stimuli-responsive nanomaterials, which react to changes in pH, temperature, light, or enzymatic environments. These materials are engineered to release drugs or change conformation only under specific biological triggers [97].

- For example, pH-sensitive polymeric nanoparticles have been developed to release anti-cancer drugs specifically within the acidic tumor microenvironment, thereby minimizing off-target effects.
- Near-infrared (NIR) light-activated gold nanorods are being explored for photothermal therapy, where controlled heating destroys cancer cells without harming nearby healthy tissue.

1.8.2 Bioinspired and Biomimetic Designs

Nature has always been the best nanotechnologist. Recent efforts in biomimicry aim to replicate the intricate surface features or functionalities of natural organisms [98].

- Nanostructured surfaces mimicking lotus leaves are being used to create antibacterial coatings by physically rupturing microbial membranes.
- Cell membrane-coated nanoparticles—cloaked in red blood cell or cancer cell membranes—can evade immune clearance and deliver therapeutics with high specificity.

1.8.3 Green and Sustainable Nanomaterial Synthesis

With rising environmental concerns, the focus has shifted to eco-friendly synthesis routes. Researchers are utilizing plant extracts, algae, and microbial systems to synthesize nanoparticles without hazardous chemicals [99].

- Silver nanoparticles synthesized using *Moringa oleifera* extract show strong antimicrobial activity and exhibit higher biocompatibility compared to chemically produced counterparts.
- Fungal-derived ZnO nanoparticles are now explored for wound healing and sunscreen formulations, offering reduced environmental burden.

1.8.4 Integrated Theranostics

The fusion of therapy and diagnostics—termed theranostics—has gained substantial traction. Here, a single nanoplatform can diagnose a disease, deliver treatment, and monitor response in real time [100, 101].

- Iron oxide nanoparticles are used as MRI contrast agents while simultaneously delivering chemotherapeutic agents.
- Quantum dots conjugated with antibodies are enabling targeted imaging and therapeutic tracking in cancer.

1.8.5 3D Nanostructured Scaffolds for Regenerative Medicine

Advances in tissue engineering have been empowered by nanotechnology. Engineered 3D nanofibrous scaffolds mimic the extracellular matrix and promote cellular adhesion, proliferation, and differentiation [102].

- Electrospun nanofibers loaded with growth factors are being utilized in nerve regeneration and bone repair.
- Hydrogel-nanocomposite scaffolds embedded with mesoporous silica nanoparticles have shown enhanced osteoconductivity in vivo.

1.8.6 AI-Driven Nanomaterial Design

Artificial intelligence (AI) and machine learning are transforming how we design nanomaterials. Predictive models now optimize compositions, surface functionalizations, and therapeutic payloads to achieve desired biological responses before lab-scale synthesis [103].

- AI-guided screening has accelerated the discovery of low-toxicity, high-efficiency nanocarriers for gene editing tools like CRISPR-Cas9.

From responsive designs and eco-friendly synthesis to AI-powered material prediction, the field of biocompatible nanomaterials stands at the crossroads of biology, technology, and sustainability. These advances are not just trends—they are testaments to a future where nanomaterials seamlessly integrate into our bodies and our biosphere, enabling precision health and regenerative possibilities never before imagined.

1.9 Challenges and Future Perspectives in Biocompatible Nanomaterials

Biocompatible nanomaterials stand at the frontier of medical innovation—tiny titans poised to revolutionize diagnostics, drug delivery, and tissue regeneration. Yet, despite their immense potential, the path forward is strewn with scientific, ethical, and translational challenges that demand urgent and interdisciplinary attention [104, 105].

1.9.1 Biological Complexity and Predictability

One of the most pressing challenges lies in the unpredictable biological interactions of nanomaterials once introduced into living systems. The complexity of biological fluids, immune responses, and cellular environments can alter the surface chemistry and biodistribution of nanoparticles.

- For instance, protein corona formation around nanoparticles can change their targeting ability, leading to reduced therapeutic efficacy or unintended immune reactions.

1.9.2 Standardization and Reproducibility

There is a glaring lack of standardized protocols for synthesis, characterization, and biocompatibility testing across laboratories. Slight variations in material size, shape, or surface charge can lead to drastically different biological outcomes, hampering reproducibility and regulatory approval.

- A study comparing identical gold nanoparticles synthesized in different labs showed up to 40% variation in cytotoxicity due to subtle synthesis differences.

1.9.3 Long-Term Toxicity and Degradation Products

While many nanomaterials show short-term safety, their long-term toxicity and the fate of their degradation products remain poorly understood. Some nanostructures persist in organs such as the liver and spleen, potentially triggering chronic inflammation or genotoxicity.

- Carbon nanotubes, for example, exhibit promising mechanical properties, but concerns about asbestos-like pulmonary effects remain unresolved.

1.9.4 Ethical and Regulatory Roadblocks

Navigating the regulatory maze for clinical translation is another significant hurdle. Nanomaterials often fall between the cracks of drug and device classifications, making their approval process ambiguous and lengthy.

- There is also growing debate on informed consent in nanomedicine trials, especially when the risks are not fully known.

1.9.5 Environmental Impact

The lifecycle impact of nanomaterials—from production to disposal—is not yet fully accounted for. Many metal-based nanoparticles such as silver or zinc oxide may accumulate in aquatic ecosystems, affecting microbial life and bioaccumulating in food chains.

- Waste from silver nanoparticle-based wound dressings can enter wastewater, where it may interfere with beneficial bacterial flora in treatment plants.

1.10 Future Perspectives

Despite these hurdles, the future of biocompatible nanomaterials is luminous, powered by synergistic advancements in material science, synthetic biology, and data-driven modeling.

1.10.1 Precision-by-Design

Future efforts will be driven by the "Precision-by-Design" approach, where predictive models and artificial intelligence guide the synthesis of nanomaterials with tailored biological outcomes.

- AI-driven simulations are now used to predict nanoparticle-cell interactions, potentially reducing reliance on animal models.

1.10.2 Personalized Nanomedicine

With the advent of genomics and personalized therapy, we foresee individualized nanocarriers engineered for specific patient profiles, diseases, and even organ environments.

- Imagine a nanoparticle that adapts its shape or surface markers dynamically based on a person's epigenetic landscape.

1.10.3 Green and Circular Nanotechnology

Sustainability will shape future nanomaterials through green synthesis and circular design, focusing on biodegradable carriers and eco-safe formulations that dissolve after use without harming biological or ecological systems.

1.10.4 Regulatory Harmonization and Ethical Frameworks

Global regulatory bodies are beginning to draft unified guidelines to streamline the path from laboratory to clinic. Ethical frameworks emphasizing transparency, patient safety, and environmental stewardship will be central to public trust and widespread adoption.

1.11 Conclusion

The journey ahead is complex but full of promise. With conscious collaboration across disciplines—chemistry, biology, engineering, ethics, and policy—we can steer the nanotechnology revolution toward a safer, smarter, and more sustainable future. The challenge is no longer just to innovate, but to innovate responsibly.

References

1. Shin M, Lim J, An J, Kim J, Lee H, Park S, Choi Y, Lee D, Kim Y, Kim S (2023) Nanomaterial-based biohybrid hydrogel in bioelectronics. Nano Convergence 10:8. https://doi.org/10.1186/s40580-023-00357-7
2. Patel J, Kumar GS, Roy H, Sharma S, Mehta A (2024) From nature to nanomedicine: bio-engineered metallic nanoparticles bridge the gap for medical applications. Discover Nano 19:85. https://doi.org/10.1186/s11671-024-04021-9
3. Wang Y, Zhang X, Yue H (2024) Two-dimensional nanomaterials induced nano-bio interfacial effects and biomedical applications in cancer treatment. J Nanobiotechnol 22:67. https://doi.org/10.1186/s12951-024-02319-5
4. Pondman K, Le Gac S, Kishore U (2023) Nanoparticle-induced immune response: health risk versus treatment opportunity? Immunobiology 228:152317. https://doi.org/10.1016/j.imbio.2022.152317
5. Mustafa G, Khan M, Ali S, Ahmad A, Iqbal M (2024) Comprehensive insights into mechanism of nanotoxicity: a focus on oxidative stress, inflammation, and immune response. Nanoscale Res Lett 19:118. https://doi.org/10.1186/s11671-024-04118-1
6. Ansari AA, Parchur AK, Li Y, Jia T, Lv R, Wang Y, Chen G (2024) Cytotoxicity and genotoxicity evaluation of chemically synthesized and functionalized upconversion nanoparticles. Coord Chem Rev 504:215672. https://doi.org/10.1016/j.ccr.2024.215672

7. Heidegger S, Goßl D, Schmidt A, Niedermayer S, Argyo C, Endres S, Bein T, Bourquin C (2016) Immune response to functionalized mesoporous silica nanoparticles for targeted drug delivery. Nanoscale 8:938–948. https://doi.org/10.1039/C5NR06122A
8. Berrecoso G, Crecente-Campo J, Alonso MJ (2024) Controlling the protein corona of polymeric nanocapsules: effect of polymer shell on protein adsorption. Drug Deliv Transl Res 14:918–933. https://doi.org/10.1007/s13346-023-01441-5
9. Singh MK, Verma AK, Yadav RK, Kumar P (2019) Surface-engineered nanoparticles for enhanced oral bioavailability of poorly water-soluble drugs. Drug Deliv Transl Res 9:675–689. https://doi.org/10.1007/s13346-019-00631-2
10. Hasany M, Taebnia N, Yaghmaei S, Shahbazi M, Mehrali M, Dolatshahi-Pirouz A, Arpanaei A (2019) Silica nanoparticle surface chemistry: an important trait affecting cellular biocompatibility in two and three dimensional culture. Colloids Surf B Biointerfaces 182:110353. https://doi.org/10.1016/j.colsurfb.2019.110353
11. Marin E (2023) History of dental biomaterials: biocompatibility, durability and still open challenges. Herit Sci 11:207. https://doi.org/10.1186/s40494-023-01046-8
12. Ghosh N, Banerjee ER (2020) Nanomaterials and their use in biomedicine. In: Banerjee ER (ed) Nanomaterials and biomedicine. Springer, Singapore, pp 51–56. https://doi.org/10.1007/978-981-15-5274-8_3
13. Sharma D, Shrivastava R, Bisht GS (2017) Nanomaterial in diverse biological applications. In: Kalia VC, Saini AK (eds) Metabolic engineering for bioactive compounds. Springer, Singapore, pp 293–317. https://doi.org/10.1007/978-981-10-5511-9_14
14. Lazurko C, Jacques E, Ahumada M, Alarcon EI (2019) Nanomaterials for its use in biomedicine: an overview. In: Alarcon EI, Ahumada M (eds) Nanoengineering materials for biomedical uses. Springer, Cham, pp 1–11. https://doi.org/10.1007/978-3-030-31261-9_1
15. Zhang L, Wang Y, Yang H, Zhao X (2020) Crossing the blood–brain barrier with nanoparticles. Drug Deliv Transl Res 10:1128–1141. https://doi.org/10.1007/s13346-020-00771-3
16. Singh S, Kumar A, Mehta R (2024) Development of biocompatible nanoparticles for targeted drug delivery and imaging applications. J Nanobiotechnol 22:123–135. https://doi.org/10.1186/s12951-024-00987-6
17. Kumari S, Sharma A, Yadav A, Yadav R (2018) Nano based drug delivery systems: recent developments and future prospects. J Nanobiotechnol 16:71. https://doi.org/10.1186/s12951-018-0392-8
18. Onzi G, Guterres SS, Pohlmann AR, Frank LA (2021) Stimuli-responsive nanocarriers for drug delivery. In: The ADME encyclopedia. Springer, Cham, pp 1–13. https://doi.org/10.1007/978-3-030-51519-5_177-1
19. Zhong JX, Raghavan P, Desai TA (2023) Harnessing biomaterials for immunomodulatory-driven tissue engineering. Regener Eng Transl Med 9:224–239. https://doi.org/10.1007/s40883-022-00279-6
20. Kurian AG, Singh RK, Sagar V et al (2024) Nanozyme-engineered hydrogels for anti-inflammation and skin regeneration. Nano-Micro Lett 16:110. https://doi.org/10.1007/s40820-024-01323-6
21. Ganguly P, Breen A, Pillai SC (2018) Toxicity of nanomaterials: exposure, pathways, assessment, and recent advances. ACS Biomater Sci Eng 4:2237–2275. https://doi.org/10.1021/acsbiomaterials.8b00068
22. Frenchibai PBB, Vinayagam R, Ambigapathi D et al (2025) Chitosan from shells of crustaceans and its application in the synthesis of biodegradable polymers. Biomed Mater Devices. https://doi.org/10.1007/s44174-025-00334-0
23. Del Prado-Audelo ML, Leyva-Gómez G, Sönmez Gürer E, Sharifi-Rad J (2023) Recent advances in modified poly(lactic acid) as tissue engineering materials. J Biol Eng 17:21. https://doi.org/10.1186/s13036-023-00338-8
24. Llamosa Peréz D, Puentes I, Murillo Romero GA, Saavedra Gaona IM, Parra Vargas CA, Rincón RJ (2024) Synthesis of superparamagnetic iron oxide nanoparticles coated with poly-

ethylene glycol as potential drug carriers for cancer treatment. J Nanopart Res 26:2. https://doi.org/10.1007/s11051-023-05900-5
25. Ariski R, Kwan Lee K, Kim Y, Lee C (2024) The impact of pH and temperature on the green gold nanoparticles preparation using Jeju Hallabong peel extract for biomedical applications. RSC Adv 14:14582–14592. https://doi.org/10.1039/D4RA00614C
26. Yanez-Aulestia A, Gupta NK, Hernández M, Osorio-Toribio G, Sánchez-González E, Guzmán-Vargas A, Rivera JL, Ibarra IA, Lima E (2022) Gold nanoparticles: current and upcoming biomedical applications in sensing, drug, and gene delivery. Chem Commun 58:10886–10895. https://doi.org/10.1039/D2CC04826D
27. Bapat RA, Chaubal TV, Joshi CP, Bapat PR, Choudhury H, Pandey M, Gorain B, Kesharwani P (2018) An overview of application of silver nanoparticles for biomaterials in dentistry. Mater Sci Eng C 91:881–898. https://doi.org/10.1016/j.msec.2018.05.069
28. Storozhuk L, Besenhard MO, Mourdikoudis S, LaGrow AP, Lees MR, Tung LD, Gavriilidis A, Thanh NTK (2021) Stable iron oxide nanoflowers with exceptional magnetic heating efficiency: simple and fast polyol synthesis. ACS Appl Mater Interfaces 13:45870–45880. https://doi.org/10.1021/acsami.1c12323
29. Wallyn J, Anton N, Vandamme TF (2019) Synthesis, principles, and properties of magnetite nanoparticles for in vivo imaging applications—a review. Pharmaceutics 11:601. https://doi.org/10.3390/pharmaceutics11110601
30. Bapat RA, Kankate PD, Vankar SG (2022) Zinc oxide (ZnO) and titanium dioxide (TiO_2) nanoparticles: biodegradable and selective cytotoxicity toward cancer cells, low toxicity in sunscreens, and use in antibacterial agents, photocatalytic cancer therapy, uv protection in cosmetics, and bone implants. J Nanomed Nanotechnol 28:1324–1333. https://doi.org/10.1016/j.jnan.2022.05.014
31. Nayak MM, Babu MD, Rajbanshi AK et al (2019) Gold–palladium bimetallic nanoparticles for photothermal therapy: synergistic effects of plasmonic heating and enhanced catalytic properties, nanomedicine: nanotechnology. Biol Med 20:102013. https://doi.org/10.1016/j.nano.2019.102013
32. Jadhav KC, Sahoo AK, Deshmukh SP et al (2019) Development of Fe–au nanoparticles for efficient drug delivery and enhanced photothermal therapy. J Nanopart Res 21:235. https://doi.org/10.1007/s11041-019-1379-9
33. Sharma BK, Tripathi RK, Sahoo PK et al (2019) Fe–au bimetallic nanoparticles: synthesis, characterization, and application in MRI contrast enhancement and photothermal therapy. J Colloid Interface Sci 548:267–274. https://doi.org/10.1016/j.jcis.2019.04.022
34. O'Connell AC, Bachilo RB, Myer CH et al (2002) Band gap fluorescence from individual single-walled carbon nanotubes. Science 297:593–596. https://doi.org/10.1126/science.1073727
35. Novoselov KS, Geim AK, Morozov SV et al (2004) Electric field effect in atomically thin carbon films. Science 306:666–669. https://doi.org/10.1126/science.1102896
36. Lazăr AI, Aghasoleimani K, Semertsidou A et al (2023) Graphene-related nanomaterials for biomedical applications. Nano 13:1092. https://doi.org/10.3390/nano13061092
37. Zarepour A, Khosravi A, Yücel Ayten N et al (2024) Innovative approaches for cancer treatment: graphene quantum dots for photodynamic and photothermal therapies. J Mater Chem B 12:4307–4334. https://doi.org/10.1039/D4TB00255E
38. Mousavi SM, Kalashgrani MY, Javanmardi N et al (2024) Recent breakthroughs in graphene quantum dot-enhanced sonodynamic and photodynamic therapy. J Mater Chem B 12:7041–7062. https://doi.org/10.1039/D4TB00767K
39. Valimukhametova AR, Fannon O, Topkiran UC et al (2024) Five near-infrared-emissive graphene quantum dots for multiplex bioimaging. Materials 11:025009. https://doi.org/10.1088/2053-1583/ad1c6e
40. Suhaan Dar M, Sahu NK (2024) Graphene quantum dot-crafted nanocomposites: shaping the future landscape of biomedical advances, discover. Nano. https://doi.org/10.1186/s11671-024-04028-2

41. Ahmed TA, Aljaeid BM (2016) Preparation, characterization, and potential application of chitosan, chitosan derivatives, and chitosan metal nanoparticles in pharmaceutical drug delivery. Drug Des Devel Ther 10:483–507. https://doi.org/10.2147/DDDT.S103068
42. Ye S, Xie C, Agar OT, Barrow CJ, Dunshea FR, Suleria HAR (2023) Alginates from Brown seaweeds as a promising natural source: a review of its properties and health benefits. J Food Sci 88:2682–2710. https://doi.org/10.1080/87559129.2023.2279583
43. Govender T, Stolnik S, Garnett MC, Illum L, Davis SS (1999) PLGA nanoparticles prepared by nanoprecipitation: drug loading and release studies of a water-soluble drug. J Control Release 57:171–185. https://doi.org/10.1016/S0168-3659(98)00239-4
44. Anbarasu M, Anandan M, Chinnasamy E, Gopinath V, Balamurugan K (2015) Synthesis and characterization of polyethylene glycol (PEG) coated Fe_3O_4 nanoparticles by chemical co-precipitation method for biomedical applications. Spectrochim Acta A Mol Biomol Spectrosc 135:536–539. https://doi.org/10.1016/j.saa.2014.07.059
45. Rethi L, Mutalik C, Anurogo D, Lu L-S, Chu H-Y, Yougbaré S, Kuo T-R, Cheng T-M, Chen F-L (2022) Lipid-based nanomaterials for drug delivery systems in breast cancer therapy. Nano 12(17):2948. https://doi.org/10.3390/nano12172948
46. Xu Y, Fourniols T, Labrak Y, Préat V, Beloqui A, Rieux A d (2022) Surface modification of lipid-based nanoparticles. ACS Nano 16(5):7168–7196. https://doi.org/10.1021/acsnano.2c02347
47. Zhao Y-Q, Li L-J, Zhou E-F, Wang J-Y, Wang Y, Guo L-M, Zhang X-X (2022) Lipid-based nanocarrier systems for drug delivery: advances and applications. Pharm Fronts 4(2):e43–e60. https://doi.org/10.1055/s-0042-1751036
48. Li X, Cheung PCK (2019) Application of natural β-glucans as biocompatible functional nanomaterials. Food Sci Human Wellness 8(4):315–319. https://doi.org/10.1016/j.fshw.2019.11.005
49. Ul-Islam M, Shehzad A, Khan S, Khattak WA, Ullah MW, Park JK (2014) Antimicrobial and biocompatible properties of nanomaterials. J Nanosci Nanotechnol 14(1):780–791. https://doi.org/10.1166/jnn.2014.8761
50. Bandara S, Du H, Carson L, Bradford D, Kommalapati R (2020) Agricultural and biomedical applications of chitosan-based nanomaterials. Nano 10(10):1903. https://doi.org/10.3390/nano10101903
51. Shukla SK, Mishra AK, Arotiba OA, Mamba BB (2013) Chitosan-based nanomaterials: a state-of-the-art review. Int J Biol Macromol 59:46–58. https://doi.org/10.1016/j.ijbiomac.2013.04.043
52. Chandrasekaran K, Tharmalingam N, Varaprasad K, Mylonakis E, Yallapu MM (2021) Biocidal and biocompatible hybrid nanomaterials from biomolecule chitosan, alginate and ZnO. Carbohydr Polym 274:118646. https://doi.org/10.1016/j.carbpol.2021.118646
53. Valizadeh A, Hosseinzadeh M, Heiran R, Hatami S, Ebrahimi MR, Bagheri M (2024) Alginate nanoparticles containing Lavandula angustifolia essential oil as a potential potent, biocompatible and low-cost antitumor agent. Polym Bull 81:1861–1874. https://doi.org/10.1007/s00289-023-04797-3
54. Elahi M, Ali S, Tahir HM, Mushtaq R, Bhatti MF (2020) Sericin and fibroin nanoparticles—natural product for cancer therapy: a comprehensive review. Crit Rev Ther Drug Carrier Syst 37(3):256–269. https://doi.org/10.1080/00914037.2019.1706515
55. Islam MS, Chen L, Sisler J, Tam KC (2018) Cellulose nanocrystal (CNC)–inorganic hybrid systems: synthesis, properties and applications. J Mater Chem B 6(6):864–883. https://doi.org/10.1039/C7TB03016A
56. Morán MC, Rosell N, Ruano G, Busquets MA, Vinardell MP (2015) Gelatin-based nanoparticles as DNA delivery systems: synthesis, physicochemical and biocompatible characterization. Colloids Surf B: Biointerfaces 134:156–168. https://doi.org/10.1016/j.colsurfb.2015.07.009
57. Staedler D, Passemard S, Magouroux T, Rogov A, Maguire CM, Mohamed BM, Schwung S, Rytz D, Jüstel T, Hwu S, Mugnier Y, Le Dantec R, Volkov Y, Gerber-Lemaire S, Prina-Mello

A, Bonacina L, Wolf J-P (2015) Cellular uptake and biocompatibility of bismuth ferrite harmonic advanced nanoparticles. Nanomedicine 11(4):815–824
58. Xie Z, Chen S, Liu Q, Lin Z, Zhao J, Fan T, Liu L, Bao S, Fan D, Zhang H (2018) Biocompatible two-dimensional titanium nanosheets for efficient plasmonic photothermal cancer therapy, arXiv preprint arXiv:1805.01976
59. Telgerd MD, Sadeghinia M, Gebremariam Birhanu MP, Daryasari AZ-K, Sadeghinia A, Hamid Akbarijavar MH, Karami ES (2019) Enhanced osteogenic differentiation of mesenchymal stem cells on metal–organic framework based on copper, zinc, and imidazole coated poly-l-lactic acid nanofiber scaffolds. J Biomed Mater Res A. https://doi.org/10.1002/jbm.a.36707
60. Fang J, Wang H, Bao X, Ni Y, Liu YTJ, Sun X, Sun Y, Li H, Zhou Y (2020) Nanodiamond as efficient peroxidase mimic against periodontal bacterial infection. Carbon 169:370–381. https://doi.org/10.1016/j.carbon.2020.07.055
61. Montalvo-Quirós S, Gómez-Graña S, Vallet-Regí M, Prados-Rosales RC, González B, Luque-Garcia JL (2021) Mesoporous silica nanoparticles containing silver as novel antimycobacterial agents against Mycobacterium tuberculosis. Colloids Surf B: Biointerfaces 197:111405. https://doi.org/10.1016/j.colsurfb.2020.111405
62. Targhi AA, Moammeri A, Jamshidifar E, Baspour KA, Sadeghi S, Lamakani L, Akbarzadeh I (2021) Synergistic effect of curcumin-Cu and curcumin-Ag nanoparticle loaded niosome: enhanced antibacterial and anti-biofilm activities. Bioorg Chem 115:105116. https://doi.org/10.1016/j.bioorg.2021.105116
63. Barmpa A, Geromichalos GD, Hatzidimitriou AG, Psomas G (2021) Nickel(II)–meclofenamate complexes: structure, in vitro and in silico DNA–and albumin–binding studies, antioxidant and anticholinergic activity. J Inorg Biochem 222:111507. https://doi.org/10.1016/j.jinorgbio.2021.111507
64. Hamdan N, Khodir WKWA, Hamid SA, Nasir MHM, Hamzah AS, Cruz-Maya I, Guarino V (2023) PCL/gelatin/graphene oxide electrospun nanofibers: effect of surface functionalization on in vitro and antibacterial response. Nano 13(3):488. https://doi.org/10.3390/nano13030488
65. Dubey R, Dutta D, Sarkar A, Chattopadhyay P (2021) Functionalized carbon nanotubes: synthesis, properties and applications in water purification, drug delivery, and material and biomedical sciences. Nanoscale Adv 3:5722–5744. https://doi.org/10.1039/d1na00293g
66. Bahrami S, Baheiraei N, Shahrezaee M (2021) Biomimetic reduced graphene oxide coated collagen scaffold for in situ bone regeneration. Sci Rep 11:16783. https://doi.org/10.1038/s41598-021-96271-1
67. Campbell E, Hasan MT, Gonzalez-Rodriguez R, Truly T, Lee BH, Green KN, Akkaraju G, Naumov AV (2021) Graphene quantum dot formulation for cancer imaging and redox-based drug delivery, nanomedicine: nanotechnology. Biol Med 37:102408. https://doi.org/10.1016/j.nano.2021.102408
68. del Castillo-Santaella T, Ortega-Oller I, Padial-Molina M, O'Valle F, Galindo-Moreno P, Jódar-Reyes AB, Peula-García JM (2019) Formulation, colloidal characterization, and in vitro biological effect of BMP-2 loaded PLGA nanoparticles for bone regeneration. Pharmaceutics 11(8):388. https://doi.org/10.3390/pharmaceutics11080388
69. Trivedi TJ, Rao KS, Kumar A (2014) Facile preparation of agarose–chitosan hybrid materials and nanocomposite ionogels using an ionic liquid via dissolution, regeneration and sol–gel transition. Green Chem 16:320–330. https://doi.org/10.1039/C3GC41317A
70. Chatterjee S, Krikorian M, Gafney HD, Gersten B (2014) Synthesis and self-assembly of lipid (DMPC)-conjugated gold nanoparticles. J Nanopart Res 16:2252. https://doi.org/10.1007/s11051-014-2252-2
71. Trivedi TJ, Rao KS, Kumar A (2011) Quick synthesis of lipid–polymer hybrid nanoparticles with low polydispersity using a single-step sonication method. Langmuir 27:14360–14365. https://doi.org/10.1021/la103576a

72. Lüdtke FL, Silva TJ, da Silva MG, Hashimoto JC, Ribeiro APB (2025) Lipid nanoparticles: formulation, production methods and characterization protocols. Foods 14:973. https://doi.org/10.3390/foods14060973
73. Dev D, Saha S, Saha S (2024) Microfluidic synthesis of lipid-based nanoparticles for drug delivery: recent advances and opportunities. Lab Chip 24:5. https://doi.org/10.1039/d3lc00821e
74. Singh AS, Singh SM, Yadav SM (2011) Chitosan nanoparticles for drug delivery. Carbohydr Polym 85:39–47. https://doi.org/10.1016/j.carbpol.2011.01.029
75. Asadian-Birjand MK, Aghaei MGR, Rahimi NM (2018) Alginate-based nanoparticles for drug delivery. Int J Biol Macromol 107:330–340. https://doi.org/10.1016/j.ijbiomac.2017.09.082
76. Gaharwar SK, Goudarzi AR, Zare HH (2014) Silk fibroin-based nanoparticles for drug delivery. Biomacromolecules 15:4070–4078. https://doi.org/10.1021/bm500886h
77. Lyu H, Gao B, He F, Ding C, Tang J, Crittenden JC (2017) Ball-milled carbon nanomaterials for energy and environmental applications. ACS Sustain Chem Eng 5:10251–10263. https://doi.org/10.1021/acssuschemeng.7b02170
78. Correard F, Maximova K, Estève M-A, Villard C, Roy M, Al-Kattan A, Sentis M, Gingras M, Kabashin AV, Braguer D (2014) Gold nanoparticles prepared by laser ablation in aqueous biocompatible solutions: assessment of safety and biological identity for nanomedicine applications. Int J Nanomedicine 9:1647–1657. https://doi.org/10.2147/IJN.S61045
79. Guo S, Chen Y-Q, Lu N-N, Wang X-Y, Xie M, Sui W-P (2014) Ultrasonication-assisted one-step self-assembly preparation of biocompatible fluorescent-magnetic nanobeads for rare cancer cell detection. Nanotechnology 25:505603. https://doi.org/10.1088/0957-4484/25/50/505603
80. Trivedi P, Gupta P, Srivastava S, Jayaganthan R, Chandra R, Roy P (2014) Characterization and in vitro biocompatibility study of Ti–Si–N nanocomposite coatings developed by using physical vapor deposition. Appl Surf Sci 293:143–150. https://doi.org/10.1016/j.apsusc.2013.12.119
81. He W, Xie Y, Xing Q, Ni P, Han Y, Dai H (2017) Sol-gel synthesis of biocompatible Eu^{3+}/Gd^{3+} co-doped calcium phosphate nanocrystals for cell bioimaging. J Lumin 192:902–909. https://doi.org/10.1016/j.jlumin.2017.08.033
82. Paduraru AV, Musuc AM, Oprea OC, Trusca R, Iordache F, Vasile BS, Andronescu E (2021) Synthesis and characterization of Photoluminescent Ce(III) and Ce(IV) substituted hydroxyapatite nanomaterials by co-precipitation method: cytotoxicity and biocompatibility evaluation. Nano 11:1911. https://doi.org/10.3390/nano11081911
83. Zare M, Namratha K, Thakur MS, Byrappa K (2019) Biocompatibility assessment and photocatalytic activity of bio-hydrothermal synthesis of ZnO nanoparticles by Thymus vulgaris leaf extract. Mater Res Bull 109:49–59. https://doi.org/10.1016/j.materresbull.2018.09.025
84. Vidal-Vidal J, Rivas J, López-Quintela MA (2006) Synthesis of monodisperse maghemite nanoparticles by the microemulsion method. Colloids Surf A Physicochem Eng Asp 288:44–51. https://doi.org/10.1016/j.colsurfa.2006.04.027
85. Chakravarty M, Sharma R, Amit K, Sharma N, Pradhan SK (2016) Synthesis and bio-compatibility study of thermal-CVD grown graphene. Int J Nanosci 15:1660016. https://doi.org/10.1142/S0219581X16600164
86. Vijayaraghavan K, Ashokkumar T (2017) Plant-mediated biosynthesis of metallic nanoparticles: a review of literature, factors affecting synthesis, characterization techniques and applications. J Environ Chem Eng 5:4866–4883. https://doi.org/10.1016/j.jece.2017.09.026
87. Bao H, Lu Z, Cui X, Qiao Y, Guo J, Anderson JM, Li CM (2010) Extracellular microbial synthesis of biocompatible CdTe quantum dots. Acta Biomater 6:3534–3541. https://doi.org/10.1016/j.actbio.2010.03.030
88. Jia H, Fan S, Cai R, Wang Z, Yuan Y, Yue T (2023) Enzyme assisted magnetic hybrids as self-activated cascade reagent with synergistic activity for antimicrobial application. Appl Surf Sci 615:156427. https://doi.org/10.1016/j.apsusc.2023.156427

89. Merlo A, Mokkapati VRSS, Pandit S, Mijakovic I (2018) Boron nitride nanomaterials: biocompatibility and bio-applications. Biomater Sci 6:2298–2311. https://doi.org/10.1039/C8BM00516H
90. Ji W, Liu J, Sha C, Yong Y-C, Jiang Y, Fang Z (2024) Nanomaterial-biological hybrid systems: advancements in solar-driven CO_2-to-chemical conversion. Green Carbon 2:322–336. https://doi.org/10.1016/j.greenca.2024.05.005
91. Jones CF, Grainger DW (2009) In vitro assessments of nanomaterial toxicity. Adv Drug Deliv Rev 61:438–456. https://doi.org/10.1016/j.addr.2009.03.005
92. Fulzele SV, Satturwar PM, Dorle AK (2003) Study of the biodegradation and in vivo biocompatibility of novel biomaterials. Eur J Pharm Sci 20:53–61. https://doi.org/10.1016/S0928-0987(03)00168-4
93. Weiss C, Diabaté S (2021) Toxicology and biocompatibility of nanomaterials. Nanomaterials 11:3110. https://doi.org/10.3390/nano11113110
94. Kumar V, Sharma N, Maitra SS (2017) In vitro and in vivo toxicity assessment of nanoparticles. Environ Chem Lett 7:243–256. https://doi.org/10.1007/s10311-017-0643-8
95. Razavi M, Khandan A (2017) Safety, regulatory issues, long-term biotoxicity, and the processing environment. Nanobiomater Sci Dev Eval:261–279. https://doi.org/10.1016/B978-0-08-100963-5.00014-8
96. Kumar P, Singh R, Kush P (2024) Biocompatibility, toxicity concerns, environmental and safety considerations, and legal aspects of functionalized magnetic nanoparticles. In: Functionalized magnetic nanoparticles for theranostic applications. Wiley, Hoboken. https://doi.org/10.1002/9781394172917.ch17
97. Mazidi Z, Javanmardi S, Naghib SM, Mohammadpour Z (2022) Smart stimuli-responsive implantable drug delivery systems for programmed and on-demand cancer treatment: an overview on the emerging materials. Chem Eng J 433:134569. https://doi.org/10.1016/j.cej.2022.134569
98. Johnson AP, Sabu C, Nivitha KP, Sankar R, Ameena Shirin VK, Henna TK, Raphey VR, Gangadharappa HV, Kotta S, Pramod K (2022) Bioinspired and biomimetic micro- and nanostructures in biomedicine. J Control Release 343:724–754. https://doi.org/10.1016/j.jconrel.2022.02.013
99. Noah NM, Ndangili PM (2022) Green synthesis of nanomaterials from sustainable materials for biosensors and drug delivery. Sens Int 3:100166. https://doi.org/10.1016/j.sintl.2022.100166
100. Singh V (2024) Theranostics: integrated diagnostics and therapy using nanomedicine. In: Nanomedicine. Springer, Berlin, pp 505–530. https://doi.org/10.1007/978-3-031-72467-1_17
101. Ni N, Zhang X, Ma Y, Yuan J, Wang D, Ma G, Dong J, Sun X (2022) Biodegradable two-dimensional nanomaterials for cancer theranostics. Coord Chem Rev 458:214415. https://doi.org/10.1016/j.ccr.2022.214415
102. Lopez de Armentia S, del Real JC, Paz E, Dunne N (2020) Advances in biodegradable 3D printed scaffolds with carbon-based nanomaterials for bone regeneration. Materials 13(22):5083. https://doi.org/10.3390/ma13225083
103. Chow JCL (2025) Nanomaterial-based molecular imaging in cancer: advances in simulation and AI integration. Biomolecules 15(3):444. https://doi.org/10.3390/biom15030444
104. Mahamuni-Badiger P, Dhanavade MJ (2023) Challenges and toxicity assessment of inorganic nanomaterials in biomedical applications: current status and future roadmaps. J Drug Deliv Sci Technol 87:104806. https://doi.org/10.1016/j.jddst.2023.104806
105. Kunzmann A, Andersson B, Thurnherr T, Krug H, Scheynius A, Fadeel B (2011) Toxicology of engineered nanomaterials: focus on biocompatibility, biodistribution and biodegradation. Biochim Biophys Acta 1810(3):361–373. https://doi.org/10.1016/j.bbagen.2010.04.007

Chapter 2
Introduction to Multifunctional Sensors

Shivarudrappa H. Pattanashetty, S. P. Nisha, and B. M. Santhosh

Abstract Multifunctional wearable systems and sensors have attracted much interest in recent years because they can monitor multiple physical, environmental, and physiological parameters simultaneously, facilitating real-time health monitoring, environmental monitoring, and smart system integration. Advances in materials science, sensor structure, and manufacturing processes have resulted in the creation of miniaturized, flexible, and robust sensors to sense temperature, pressure, strain, humidity, light, gas, and acceleration—frequently in a single device. These sensors take advantage of novel materials, like low-dimensional nanostructures and upconverting nanoparticles, and new integration schemes like planar, 3D, and stacked structures. In addition, the integration of sensor fusion, energy harvesting, and information storage capability in multifunctional platforms gives rise to new opportunities for wearable electronics, human–machine interfaces, and Internet of Things (IoT) applications. In spite of fault tolerance and long-term reliability challenges, ongoing design optimization and innovative fabrication technologies are propelling the field toward more flexible, capable, and affordable multifunctional sensing solutions.

Keywords Sensors · Electrochemical properties · Nanoparticles

2.1 Introduction

The fast pace of digital technology and smart system evolution has created an increasing need for intelligent, integrated, and small-sized sensing solutions. Multifunctional sensors have been key elements in solving today's challenges in healthcare, environmental monitoring, robotics, and consumer electronics. These sensors are specifically designed to detect various physical, chemical, or biological

S. H. Pattanashetty · S. P. Nisha · B. M. Santhosh (✉)
Department of Chemistry, G M University, Davanagere, Karnataka, India

R. Naik et al. (eds.), *Biocompatible Nanomaterials for Smart Sensors*, Nanostructure Science and Technology,
https://doi.org/10.1007/978-3-032-01745-1_2

parameters simultaneously, greatly enhancing the efficiency of the system, minimizing cost and volume, and increasing the number of potential applications. The historical path of sensor technology demonstrates the distinct shift away from single-function transducers toward very highly integrated multifunctional platforms. This development is not simply a technological nicety but one that is necessitated by the rising complexity of human environments and the systems that engage with them. The vision of a context-aware, real-time, and seamless interface between the physical and digital worlds necessitates sensors not only sensitive and selective but also compact, durable, and operational under different stimuli and environmental conditions.

In wearable electronics, for example, multifunctional sensors are critical to improving personal health monitoring systems. Early wearable sensors go back as far as the 1960s in space missions, where heart rates and body temperatures of astronauts were detected. These early systems formed the basis of advanced wearable devices that nowadays provide more complex functions such as pulse wave analysis, gait recognition, and multimodal health data measurement. Due to advances in flexible and stretchable electronics, today's wearable sensors are skin-conformable, lightweight, and can withstand repetitive mechanical stress without performance degradation [1].

Apart from wearables, the increased demand for the Internet of Things (IoT) and wireless sensor networks (WSNs) has further emphasized the requirement for multifunctional sensors integrated together. These networks focus on building connected systems that are able to capture, analyze, and respond to environmental and physiological information in real-time. As conceived by the "Smart Dust" and "CeNSE" paradigms, sensors integrated into pervasive platforms such as mobile phones, environmental nodes, or even infrastructure components are the digital nervous system of contemporary ecosystems. In such cases, multifunctionality becomes essential to reduce power usage, size, and cost to achieve long-term, low-maintenance deployments [2].

Multifunctional sensors, by the very name, combine two or more sensing modalities—e.g., temperature, pressure, humidity, light, gas, strain, or bioelectrical signals—into one device or system. By this combination, a number of advantages are achieved: smaller footprint, better data correlation and context-awareness, reduced production and maintenance costs, and improved system reliability. But it presents engineering challenges as well. Balancing performance trade-offs among sensing components, cross-interference avoidance, structural robustness, and calibration accuracy across modalities are main concerns. Materials science has been a game-changing force for addressing these challenges. Low-dimensional materials like 0D nanoparticles, 1D nanowires, and 2D nanosheets have been especially useful for the design of multifunctional sensors based on their high surface area, mechanical flexibility, and electrical, optical, and thermal properties, which can be readily tuned. These materials can be incorporated into polymer matrices or structured in hybrid architectures, facilitating various transduction mechanisms like piezoresistive, capacitive, thermoelectric, and piezo-phototronic effects. For example, carbon black, silver nanoparticle, or quantum dot-based nanocomposites have shown excellent sensitivity toward strain, pressure, and temperature simultaneously, in addition to enabling optical or energy-harvesting functionalities [3]. Sometimes multifunctional sensors also possess energy-harvesting ability and transform ambient energy (mechanical, thermal, or optical) into electrical power to energize themselves or the surrounding circuitry. They become the best choice for remote, battery-less, or long-duration applications. Having such functionality embedded in a common platform not only enhances autonomy but also supports sustainable and scalable design paradigms necessary for large-scale implementation in smart cities, medicine, and environmental protection. Device structure and system integration design principles also affect sensor performance. Technologies such as planar integration, stacked 3D geometries, and folded or origami-type configurations provide high surface area and complex signal capture in the smallest possible footprint. High-reliability signal processing algorithms, especially in multi-sensor data fusion, commonly aid such structures where autonomous signals from multiple modalities are integrated to produce a cohesive picture of a system or environment. Such integration improves precision, minimizes false alarms, and offers multidimensional information crucial in areas such as autonomous robotics or clinical diagnosis [4]. A real-world example of such integration is seen with the MFISEs (Multi-Functional Integrated Sensors for the Environment) platform, which integrates 10 various sensing functions—such as acceleration (3-axis), temperature, humidity, airspeed, etc.—on a single silicon die. This degree of sensor fusion not only minimizes component count but also streamlines signal routing, increases reliability, and provides enormous benefits in power, space, and cost savings. The fabrication approach adopted in MFISEs embodies the wider trend in sensor engineering—away from discrete sensor assemblies and toward monolithic integration with microelectromechanical systems (MEMS) technologies [2]. In a similar manner, lanthanide-doped upconversion multifunctional optical sensors offer a refined means of remote, ratiometric measurement of parameters like temperature, vacuum pressure, and laser excitation power density. Utilizing distinct luminescence features, these systems provide noncontact high-resolution monitoring with new prospects in situations where electrical contacts are impractical or dangerous [5]. Although all thse advancements are made, there are still some limitations present. These include reliability over extended durations under dynamic load, bonding rigid components with compliant substrates, cross-sensitivity between the different sensing modes, and scaling the manufacturing. A multidisciplinary solution by involving materials engineering, microfabrication, circuitry design, and data science solves these problems. Future directions include machine-learning-enabled sensor calibration, self-healing electronics, biodegradable devices, and application-specific sensor platforms that are modular and adjustable for user needs.

Furthermore, user-centric design is gaining more significance. Comfort, esthetics, and user interface design become important for mass acceptance in healthcare and consumer electronics. Multifunctional sensors should be discreet, simple to use, and allow wireless interaction with smartphones or cloud systems. Printed electronics and modulated e-tattoo platforms are already trending in this direction, promising next-generation devices that are user-friendly and high-performing [1]. Finally, multisensing systems are a pillar of tomorrow's intelligent technologies. Their potential to integrate several sensing functions onto a single platform enables a broad range of applications from continuous health monitoring and environmental monitoring to factory automation and smart infrastructure. As material properties, integration methods, and data interpretation algorithms continue to be refined, these sensors will increasingly contribute to the development of adaptive, efficient, and intelligent systems in many fields.

2.2 Material Selection for Multifunctional Sensors

The functionality and performance of multifunctional sensors are highly reliant on the properties of the materials used to build them. A perfect material for such sensors should have high sensitivity toward a variety of external stimuli ranging from pressure, temperature, strain, and light and be mechanically flexible to support wearable and conformal applications [6]. Additionally, biocompatibility is required in skin-contact and biomedical applications, and environmental stability is important so that it maintains reliability under environmental conditions like humidity, UV, and oxidation. Most importantly, the material would need to promote multifunctionality, i.e., it has the ability to transduce a variety of types of signals across one platform.

Material science, thus, is at the forefront of developing these sensor systems. Low-dimensional nanomaterials, consisting of 0D (nanoparticles), 1D (nanowires and nanotubes), and 2D (nanosheets) structures, are among the most promising candidates. These materials possess tunable and unique electrical, thermal, and optical properties that greatly improve sensor performance [7].

Notably, 0D nanoparticles and quantum dots like silver (Ag), gold (Au), titanium dioxide (TiO_2), and carbon dots are commonly employed as fillers within composite sensor platforms. They exhibit sensitivity to mechanical deformation, heat, and light and alter electrical conductivity pathways in response, which is vital for piezoresistive and capacitive sensing. Analogously, 1D nanomaterials such as carbon nanotubes (CNTs) and ZnO nanowires are incorporated into piezotronic and piezophototronic sensors, wherein mechanical stress affects their electronic properties directly. Last but not least, 2D nanosheets including graphene, MXenes, and molybdenum disulfide (MoS_2) introduce high surface sensitivity, good charge mobility, and superior mechanical strength. Coupling them into flexible substrates facilitates the realization of sensors capable of sensing strain, gases, and temperature variations in a concurrent manner.

Overall, the strategic choosing and utilization of these sophisticated materials form the backbone of the engineering of next-generation multifunctional sensors with a high level of sensitivity, adaptability, and longevity in numerous real-world environments.

Materials used as sensors tend to be fabricated through the incorporation of nanomaterials in several different polymer matrices, inorganic semiconductors, and composites in order to advance functionality and versatility. Elastomer polymers, e.g., PDMS, PU, and Ecoflex, possess a great extent of stretchability as well as mechanical strength and hence can serve well as substrates in wearables. Hydrogels are ideally suited for biosensing and temperature sensing based on their high water content and ionic conductivity. Conductive polymers such as polyaniline (PANI), polypyrrole (PPy), and PEDOT:PSS are inherently conductive and are widely used in strain or electrochemical sensors. Inorganic semiconductors such as silicon (Si), gallium nitride (GaN), and zinc oxide (ZnO) are typically combined with MEMS technologies to create piezoresistive, optical, and pressure sensors. Lanthanide-doped oxides such as $NaYF_4$:Yb,Er are being widely utilized for luminescent thermometry and power density sensing owing to their characteristic optical properties. In addition, composite materials of conductive fillers mixed with polymers allow for the development of multifunctional sensors to measure pressure, temperature, and strain in one instance. The preparation of these sensor materials utilizes a number of advanced methods. In the case of nanoparticles, wet chemical approaches like sol-gel, reduction, and co-precipitation are used to synthesize different metal oxides and nanoparticles. Hydrothermal and solvothermal synthesis produce well-crystallized nanostructures such as ZnO nanowires and MXenes, whereas microwave-assisted synthesis enables rapid, uniform nanoparticle synthesis. Conductive polymers are prepared through chemical oxidative polymerization or electrochemical polymerization, providing thickness and doping control—essential for sensor electrode design. Nanocomposites are formulated by in situ polymerization, ultrasonication blending and casting, or highly controlled layer-by-layer (LbL) assembly based on electrostatic interactions, which offer engineered properties and structural control for demanding sensing applications [8].

2.3 Fabrication of Multifunctional Sensors

Fabrication of multifunctional sensors demands a delicate trade-off among sensitivity, power requirement, spatial density, and manufacturing convenience. A number of integration schemes are used to address these requirements. One involves the utilization of single materials with intrinsic multifunctionality—capitalizing on properties like piezoresistivity, thermal conductivity, and optical responsivity to sense several environmental variables in parallel. Another method is planar integration, in which several sensing elements are placed laterally on rigid or flexible substrates. More sophisticated is three-dimensional (3D) integration, in which sensors are stacked vertically to maximize space efficiency and functional density.

The fabrication method of choice is determined by design needs and the application. Standard fabrication methods involve MEMS processes like deep reactive ion etching (DRIE), diffusion doping, and thin-film deposition, particularly for silicon-based integrated sensors. For flexible or stretchable systems, nanomaterials—such as silver or gold nanoparticles and carbon nanotubes (CNTs)—are embedded inside elastomeric matrices like PDMS or TPU, which results in conductive, deformable sensing layers. Mechanically guided 3D assembly enables flat 2D precursors to be reconfigured into intricate 3D geometries, which are suitable for wearable sensors.

Advanced sensor platforms, such as Multi-Functional Integrated Sensors for the Environment (MFISE), demonstrate the integration of several sensing capabilities—temperature, humidity, airspeed, and acceleration—into a single chip with up to five photolithography masks. Such integration mirrors existing trends toward miniaturizing and making sensors multifunctional in mobile and embedded systems [9].

Fabrication is initiated by substrate preparation, where substrates like PDMS, PET, polyimide (Kapton), or fabric-based substrates are used. Active materials are deposited using spin coating, drop casting, blade coating, or inkjet or 3D printing techniques. This is then followed by curing of conductive composites with heat or UV light [2].

Patterning is essential to sensor accuracy and involves photolithography for micro-scale structuring, screen printing for economically viable large-area electronics, laser ablation or laser-induced graphene (LIG) for in-place electrode creation, and soft lithography for micro-patterning of flexible substrates.

Device integration is next, where sensors are mounted with electrodes through evaporation, printing, or embedding. In 3D arrangements, stacked layering with insulating barriers enables the integration of different sensing modalities. Electrical connection is made through wire bonding, silver paste, anisotropic conductive films (ACF), or soldering.

To ensure stability and durability, encapsulation methods like parylene coating, lamination films, or epoxy resins offer biocompatibility and environmental protection. Lastly, strict calibration and testing using temperature-controlled chambers, pressure rigs, or electrochemical stations guarantee accuracy (Table 2.1; Graph 2.1).

Table 2.1 Summary of different materials commonly employed in multifunctional sensors, emphasizing their synthesis methods and corresponding sensing functionalities

Material type	Representative materials	Typical synthesis methods	Primary sensing applications
0D nanomaterials	Ag, Au, TiO_2, carbon dots	Wet chemical reduction, sol-gel	Optical sensing, pressure, temperature
1D nanomaterials	Carbon nanotubes (CNTs), ZnO wires	Chemical vapor deposition, hydrothermal	Strain, pressure, piezoresistive sensing
2D nanomaterials	Graphene, MXenes, MoS_2	Etching and delamination, solution exfoliation	Gas detection, pressure, electrical, and optical sensing
Conductive polymers	PEDOT:PSS, polyaniline (PANI), PPy	Chemical/ electrochemical polymerization	Electrochemical, strain, biosensing
Inorganic semiconductors	ZnO, Si, GaN, $NaYF_4$:Yb,Er	Hydrothermal, sol-gel, co-precipitation	Thermometry, photodetection, pressure (vacuum)
Nanocomposites	CNT/PDMS, graphene/PU, AgNP/ hydrogel	In situ polymerization, blending, casting	Strain, temperature, pressure, gas

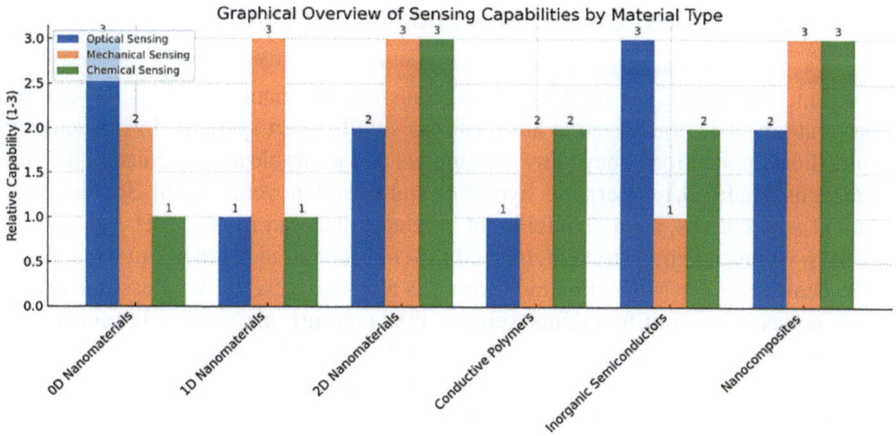

Graph 2.1 Relating the sensing proficiencies of different material types across three key domains: optical, mechanical, and chemical

Structural and electrical characterization techniques like SEM, TEM, XRD, FTIR/Raman spectroscopy, and electrical measurements (IV/CV) confirm the device's composition, morphology, and functionality [10].

2.4 Signal Processing and Sensor Fusion Techniques

Signal reconstruction and sensor fusion are central to multifunctional sensor system performance. Every sensor channel produces a data stream, usually corrupted by noise or environmental interference [11]. Proper signal reconstruction is critical to converting raw measurements into trustworthy and interpretable parameters. Multivariate transfer functions and linear, multiple linear, and polynomial regression models are standardly applied to map sensor outputs onto target input variables [12]. More sophisticated methods such as total least squares (TLS), Moore-Penrose pseudoinverses, and Kalman Filtering are utilized to deal with noisy or ill-conditioned data matrices. Multifunctional sensors tend to be cross-sensitive, meaning a sensor's reaction to one stimulus influences another measurement [13]. Although normally regarded as a limitation, this property is used strategically to estimate several parameters from a single stream of signals—for example, using the thermal response of a material to measure pressure and temperature simultaneously [14]. Artificial intelligence-based models such as General Regression Neural Networks (GRNN) and Radial Basis Function Neural Networks (RBFNN) have been proposed to tackle the signal mapping and error correction, providing a path toward digital sensor readout and improved fault tolerance [15].

2.5 Applications, Challenges, and Future Outlook

Multifunction sensors are on the verge of transforming many industries through the combination of various sensing abilities into small, smart systems. In medical and biomedical applications, they allow continuous and wearable measurement of critical vital signs like ECG, temperature, hydration status, and movement [16]. Sophisticated clinical applications, such as intelligent bandages that can monitor pH and glucose levels in wounds, demonstrate their capability for patient-centered monitoring. Aside from healthcare, environmental monitoring is also enhanced by wireless sensor network nodes with multifunctional sensors that measure air quality, humidity, and atmospheric pressure, improving real-time environmental evaluation [17].

In robotics and smart systems, multifunctional sensors give electronic skins, prosthetics, and robotic surfaces human-like touch perception, allowing for sophisticated interaction with the environment and people. Likewise, industrial systems tap capacitive and admittance-type sensors to track fluid levels, pressure, and temperature in tanks to maximize operational efficiency and safety [18].

But implementation of multifunctional sensors poses engineering difficulties. Scaling up the number of sensing elements can degrade system reliability and increase the risk of failure, and thus redundancy, self-diagnosis, and AI-based predictive maintenance practices have to be employed to guarantee sound performance [19]. Furthermore, miniaturization typically comes with trade-offs, which may reduce sensor sensitivity and selectivity unless counteracted by novel materials or signal processing methods [20]. Lastly, controlling cross-sensitivity—where sensor channels interfere or synergize—calls for strategic material choice and circuit design to preserve measurement integrity and functional consonance.

Overall, the intersection of multifunctionality, intelligent integration, and strong design principles makes them integral to the future generation of smart, adaptive systems.

2.6 Conclusion

The area of multifunctional sensors has witnessed tremendous growth in the last few decades due to an ever-growing need for smart, miniaturized, and integrated sensing devices in many fields like healthcare, environmental monitoring, industrial automation, and wearable electronics. The current chapter offered an introductory yet exhaustive summary of multifunctional sensors, explaining their basic working principles, types of stimuli that they sense, materials used for their fabrication, and their vast array of applications in the real world.

Multifunctional sensors are distinguished by the fact that they can sense and react to several physical, chemical, or biological stimuli under single or sequential input. Not only does this make the devices more efficient and space-saving, it also decreases the number of single discrete sensors needed. With the increasing trend toward the Internet of Things (IoT), smart cities, individualized healthcare, and intelligent environmental monitoring systems, the use of multifunctional sensors has become more important than ever before. These sensors provide real-time data acquisition and analysis, enabling timely decision-making and automation, which are essential in today's technology-driven and fast-paced world. One of the main reasons for the development of multifunctional sensors is the evolution in material science and nanotechnology. Nanomaterials like carbon nanotubes, graphene, MXenes, transition metal dichalcogenides (TMDs), metal-organic frameworks (MOFs), and perovskites have revolutionized the design of sensors. Their high surface area, good electrical conductivity, chemical functionality tunability, and flexibility on compatible substrates have made them excellent candidates to serve as next-generation sensor platforms. These materials make sensors superior in terms of sensitivity, selectivity, and stability, even in harsh environments.

Additionally, the combination of several sensing mechanisms—piezoelectric, pyroelectric, electrochemical, and optical transduction—into one platform has enabled the creation of sensors that can exhibit multifunctionality at both the hardware and software levels. With the introduction of hybrid sensing systems that integrate various transduction principles and data fusion algorithms, the functionality and robustness of sensor systems are further augmented. Further, with the integration of energy harvesting technologies like triboelectric and piezoelectric nanogenerators, self-powered sensors are developed, and external power sources are avoided, enhancing the sustainability of devices. Multifunctional sensors also take a central role in biomedical and healthcare monitoring applications. Wearable sensors that can track physiological signals like body temperature, pulse rate, sweat chemistry, and movement are facilitating personal medicine and ongoing health monitoring. The sensors have proved especially useful during times of public health emergency like the COVID-19 pandemic, when real-time monitoring and remote diagnosis were crucial. In the same vein, in environmental monitoring, multifunctional sensors are utilized to identify pollutants, toxic gases, humidity, and temperature fluctuations, helping to develop more sustainable and safer environments.

In the future, a number of issues have to be resolved in order to leverage the full potential of multifunctional sensors. These are long-term stability, biocompatibility for implantable sensors, cross-sensitivity of different stimuli, and integration with flexible and stretchable substrates. Also, mass-scale production with cost-effectiveness has yet to be achieved. In addition, advanced processing techniques based on machine learning and artificial intelligence are becoming increasingly merged with sensor signals to translate into better interpretability, predictability, and automation.

In summary, multifunctional sensors are destined to become an increasingly important driver of future smart technology and networked systems. Their interdisciplinary approach, which brings physics, chemistry, materials science, and electronics together, emphasizes their versatility across a wide spectrum of applications. On-going development of materials, design structures, and signal processing will lead to progressively more resilient, intelligent, and energy-frugal sensor systems. As we move toward an increasingly globalized and automated society, multifunctional sensors will surely be at the helm of technological progress.

References

1. Jia S, Gao H, Xue Z, Meng X (2022) Recent advances in multifunctional wearable sensors and systems: design, fabrication, and applications. Biosensors 12(11):1057
2. Roozeboom CL, Sim JY, Wickeraad D, Dura B, Smith WS, Hopcroft MA, Pruitt BL (2012) Multi-functional integrated sensors for the environment. In: 2012 IEEE 25th international conference on micro electro mechanical systems (MEMS). IEEE, Piscataway, pp 144–147
3. Xu Q, Dai Y, Peng Y, Hong L, Yang N, Wang Z (2021) Recent development of multifunctional sensors based on low-dimensional materials. Sensors 21(22):7727
4. Majumder BD, Roy JK, Padhee S (2018) Recent advances in multifunctional sensing technology on a perspective of multi-sensor system: a review. IEEE Sensors J 19(4):1204–1214
5. Hernández-Álvarez C, Brito-Santos G, Martín IR, Sanchiz J, Saidi K, Soler-Carracedo K, Runowski M (2023) Multifunctional optical sensing platform of temperature, pressure (vacuum) and laser power density: NaYF 4: Gd 3+, Yb 3+, Er 3+ nanomaterial as luminescent thermometer, manometer and power meter. J Mater Chem C 11(30):10221–10229
6. Shieh J, Huber JE, Fleck NA, Ashby MF (2001) The selection of sensors. Prog Mater Sci 46(3–4):461–504
7. Begley MR (2005) The impact of materials selection and geometry on multi-functional bilayer micro-sensors and actuators. J Micromech Microeng 15(12):2379
8. Noah NM (2020) Design and synthesis of nanostructured materials for sensor applications. J Nanomater 2020(1):8855321
9. Xie M, Hisano K, Zhu M, Toyoshi T, Pan M, Okada S, Bowen C (2019) Flexible multifunctional sensors for wearable and robotic applications. Adv Mater Technol 4(3):1800626
10. Guo Y, Wei X, Gao S, Yue W, Li Y, Shen G (2021) Recent advances in carbon material-based multifunctional sensors and their applications in electronic skin systems. Adv Funct Mater 31(40):2104288
11. Luo RC, Yih CC, Su KL (2002) Multisensor fusion and integration: approaches, applications, and future research directions. IEEE Sensors J 2(2):107–119
12. Kong L, Peng X, Chen Y, Wang P, Xu M (2020) Multi-sensor measurement and data fusion technology for manufacturing process monitoring: a literature review. Int J Extreme Manuf 2(2):022001
13. Mitchell HB (2007) Multi-sensor data fusion: an introduction. Springer, Cham
14. Wu J, Su Y, Cheng Y, Shao X, Deng C, Liu C (2018) Multi-sensor information fusion for remaining useful life prediction of machining tools by adaptive network based fuzzy inference system. Appl Soft Comput 68:13–23
15. Servati A, Zou L, Wang ZJ, Ko F, Servati P (2017) Novel flexible wearable sensor materials and signal processing for vital sign and human activity monitoring. Sensors 17(7):1622
16. Zhang Y, Ding J, Qi B, Tao W, Wang J, Zhao C, Shi J (2019) Multifunctional fibers to shape future biomedical devices. Adv Funct Mater 29(34):1902834
17. Mamun MAA, Yuce MR (2020) Recent progress in nanomaterial enabled chemical sensors for wearable environmental monitoring applications. Adv Funct Mater 30(51):2005703
18. Moinudeen GK, Ahmad F, Kumar D, Al-Douri Y, Ahmad S (2017) IoT applications in future foreseen guided by engineered nanomaterials and printed intelligence technologies a technology review. Int J Internet Things 6(3):106–148
19. Ferreira ADB, Nóvoa PR, Marques AT (2016) Multifunctional material systems: a state-of-the-art review. Compos Struct 151:3–35
20. Adday GH, Subramaniam SK, Zukarnain ZA, Samian N (2022) Fault tolerance structures in wireless sensor networks (WSNs): survey, classification, and future directions. Sensors 22(16):6041

Chapter 3
Synthesis of Biocompatible Nanomaterials

S. Ishwarya, Ganesh Koyyada, Burragoni Sravanthi Goud, Ongera Gilbert, and Krishna Prakash Arunachalam

Abstract Innovation in biomedical domains such drug delivery, imaging, diagnostics, and tissue regeneration is largely driven by biocompatible nanomaterials. Their synthesis necessitates striking a delicate balance between assuring safe biological interactions and obtaining ideal material qualities. This chapter provides a thorough summary of the main synthesis methods for biocompatible nanomaterials, including physical, chemical, and environmentally acceptable green approaches. Every strategy is evaluated according to its impact on the biocompatibility of the material, cost, scalability, and effectiveness. Green synthesis, which uses natural resources including extracts from plants and microorganisms to create sustainable and nontoxic nanomaterials, is given particular attention. Additionally, the significance of surface modification in enhancing functionalization potential and biological compatibility is examined. Along with addressing important issues including toxicity, production uniformity, and regulatory problems, the review also highlights potential future paths and cutting-edge technologies that could improve synthesis procedures and hasten their use in clinical settings. All things considered, this paper acts as a manual for researchers aiming to create safe and effective nanomaterials for use in biomedicine.

Keywords Green synthesis · Nanomaterials · Biocompatibility

S. Ishwarya (✉)
Department of Studies in Chemistry, Davangere University, Davangere, Karnataka, India

G. Koyyada · B. S. Goud
School of Chemical Engineering, Yeungnam University, Gyeongsan, South Korea

O. Gilbert
Department of Mathematics, Actuarial and Physical Sciences, University of Kabianga, Kericho, Kenya

K. P. Arunachalam
Departamento de Ciencias de la Construcción, Facultad de Ciencias de la Construcción Ordenamiento Territorial, Universidad Tecnológica Metropolitana, Santiago, Chile

R. Naik et al. (eds.), *Biocompatible Nanomaterials for Smart Sensors*, Nanostructure Science and Technology,
https://doi.org/10.1007/978-3-032-01745-1_3

3.1 Introduction

Nanotechnology has ushered in a new era in material science, offering tools to engineer materials at the nanoscale level typically between 1 and 100 nm in at least one dimension [1–3]. These nanomaterials exhibit remarkable characteristics, such as high surface area, enhanced reactivity, and novel optical, magnetic, or mechanical behaviors not found in bulk materials. As a result, they are increasingly applied in diverse sectors, including electronics, energy, medicine, and environmental science. Among these fields, biomedicine stands out as one of the most promising areas for nanomaterial application. However, their integration into biological systems hinges on a key factor: biocompatibility. In simple terms, biocompatibility is the ability of a material to function appropriately in contact with living tissue without provoking harmful effects, such as toxicity or immune response [4]. For nanomaterials, achieving biocompatibility is a complex but essential challenge, as their interaction with cells, tissues, and organs determines their success in clinical or environmental contexts. Biocompatible nanomaterials are extensively explored in drug delivery, where they can carry therapeutic agents precisely to target sites, thereby minimizing side effects and improving drug efficacy. Similarly, in diagnostics, nanoparticles are used to enhance contrast in imaging techniques like MRI and CT scans, improving disease detection. In tissue engineering, nanostructured scaffolds promote cellular growth and repair by mimicking the natural extracellular matrix. These functions, however, require that the materials neither interfere with biological processes nor accumulate in harmful ways. In environmental applications, biocompatibility is equally vital. Nanomaterials are now widely used in pollution control, water purification, and environmental sensing. Materials that degrade safely or are nontoxic to ecosystems are preferred to avoid long-term environmental damage. If improperly designed, nanomaterials can become pollutants themselves, posing risks to wildlife and potentially entering the food chain. The ability of a nanomaterial to behave safely and effectively in these applications is deeply influenced by how it is synthesized. Synthesis determines critical properties such as particle size, morphology, surface charge, and chemical composition—all of which directly impact biological interactions. For example, conventional chemical synthesis methods can yield well-defined nanostructures but may leave behind toxic residues or require hazardous reagents. In contrast, green synthesis approaches—using biological agents such as plant extracts, bacteria, fungi, or algae—offer a more environmentally sustainable and often more biocompatible alternative [5, 6]. Physical synthesis methods, including laser ablation, evaporation–condensation, and high-energy milling, can produce highly pure and crystalline nanoparticles. However, these techniques are often energy-intensive and lack fine control over surface properties, which are essential for biological compatibility. Moreover, they frequently require additional chemical modifications post-synthesis to improve stability and functionalization. Chemical methods, such as sol-gel processing, co-precipitation, and hydrothermal synthesis, are widely adopted due to their reproducibility, scalability, and ability to control particle morphology. Yet, these methods can introduce contaminants or involve harsh conditions that reduce the biocompatibility of the final product unless carefully optimized [7]. To overcome these limitations, researchers often apply surface modification strategies. Coating nanoparticles with biocompatible substances—such as polyethylene glycol (PEG), chitosan, or other polymers can shield them from immune recognition and extend their circulation time in the body. Additionally, surface functionalization with targeting molecules enables site-specific delivery, reducing off-target interactions and side effects. Given these factors, the choice of synthesis method is critical. It not only influences the physical and chemical properties of nanomaterials but also defines their safety profile and biological performance. Thus, selecting an appropriate synthesis route is central to the development of nanomaterials for sensitive applications in healthcare and the environment [8].

3.2 Classification of Biocompatible Nanomaterials

The classification of biocompatible nanomaterials can be approached from various perspectives, including their composition, origin, and structure/morphology. These classifications help define their biological interactions, functionalities, and potential applications. Understanding these categories is essential for selecting the right type of nanomaterial for specific biomedical and environmental applications [9].

3.2.1 Based on Composition

Biocompatible nanomaterials can be categorized based on their chemical composition. The material composition directly affects its biological behavior, such as cellular uptake, toxicity, biodegradability, and interaction with biomolecules. Broadly, biocompatible nanomaterials can be grouped into several categories based on the materials used in their fabrication [10–13]:

Metal-Based Nanomaterials Metal nanoparticles such as gold (Au), silver (Ag), and iron oxide (Fe_3O_4) are widely studied due to their tunable properties, ease of synthesis, and biocompatibility in various biomedical applications. Gold nanoparticles, in particular, are widely used in imaging, diagnostics, and drug delivery due to their excellent optical properties and low toxicity. Silver nanoparticles are recognized for their antimicrobial properties, while iron oxide nanoparticles are frequently utilized for magnetic resonance imaging (MRI) and targeted drug delivery.

Polymeric Nanomaterials Polymeric nanoparticles are highly versatile and often employed in controlled drug delivery systems. Materials such as poly (lactic-co-glycolic acid) (PLGA), polyethylene glycol (PEG), and chitosan are popular due to their biodegradability, low toxicity, and ease of functionalization. Polymers can be used to encapsulate drugs, proteins, or nucleic acids, providing protection and controlled release, which is crucial for therapeutic applications.

Lipid-Based Nanomaterials Lipid-based nanoparticles, such as liposomes and solid lipid nanoparticles, are used extensively in drug delivery and gene therapy. Their structure, which closely resembles cellular membranes, facilitates the delivery of both hydrophilic and hydrophobic drugs. Lipid nanoparticles are also used in mRNA vaccine delivery due to their ability to protect sensitive therapeutic agents and promote cellular uptake.

Carbon-Based Nanomaterials Carbon-based nanomaterials, including carbon nanotubes (CNTs), graphene, and fullerenes, have gained attention for their unique electronic properties, mechanical strength, and biocompatibility. However, the toxicity of some forms of carbon nanotubes and their potential to accumulate in tissues have raised concerns. Surface modification of carbon-based materials is often employed to enhance their biocompatibility and reduce their toxicity.

Ceramic Nanomaterials Inorganic materials such as silica nanoparticles, hydroxyapatite, and calcium phosphate nanoparticles are commonly used in biomedical applications due to their biocompatibility and ability to promote osteointegration in bone tissue engineering. Silica nanoparticles, for instance, are widely used for drug delivery and imaging due to their porous structure and ease of functionalization.

Hybrid Nanomaterials Hybrid nanomaterials combine two or more materials with complementary properties, such as organic and inorganic components. These hybrids leverage the advantages of both types of materials, offering enhanced stability, multifunctionality, and biocompatibility. Examples include gold-silica nanoparticles or polymer-metal composites, which are used in applications such as targeted drug delivery and diagnostic imaging.

3.2.2 Based on Origin

Biocompatible nanomaterials can also be classified according to their origin, whether they are natural or synthetic. Each type has its advantages and potential limitations in terms of safety, sustainability, and performance in biological systems.

Natural Nanomaterials Natural nanomaterials are derived from biological sources, such as plants, animals, or microorganisms. These materials are inherently biocompatible and biodegradable, making them attractive for biomedical and environmental applications. Common examples include plant-derived nanoparticles, fungal nanoparticles, and bacterial nanocellulose. Plant-based nanoparticles, for instance, are produced by using plant extracts as reducing agents in the synthesis process, often yielding nanoparticles with low toxicity. Chitosan, a natural polymer derived from crustacean shells, is widely used for drug delivery systems due to its biodegradability and nontoxic nature [14].

Synthetic Nanomaterials Synthetic nanomaterials are engineered through chemical or physical methods and can be tailored to exhibit specific properties. While they offer a high degree of control over their size, shape, and surface chemistry, their biocompatibility may need to be enhanced through surface modification or functionalization. Synthetic nanomaterials such as polymeric nanoparticles, metal oxide nanoparticles, and carbon nanotubes are commonly used in a variety of applications. One challenge with synthetic materials is ensuring that they are biocompatible and do not trigger unwanted immune responses or toxicity in the body [15].

Biomimetic Nanomaterials Biomimetic nanomaterials are a subset of synthetic nanomaterials that are designed to imitate the structures or functions of natural materials. These materials are created to mimic the biological behavior of naturally occurring substances like proteins, lipids, and polysaccharides. Biomimetic nanomaterials aim to combine the advantages of both natural and synthetic materials, such as enhanced biocompatibility and customized functionalities. They are gaining traction in applications such as tissue engineering, where they can support cell growth and tissue regeneration by mimicking the extracellular matrix [16].

3.2.3 Based on Structure and Morphology

The structure and morphology of nanomaterials significantly affect their biocompatibility, interaction with biological systems, and their potential for targeted delivery and imaging applications. Based on their structural features, nanomaterials can be classified as follows [17–20].

Zero-Dimensional (0D) Nanomaterials Zero-dimensional nanomaterials, such as quantum dots and nanoparticles, have no extended dimensions beyond their size (typically spherical). They are often used in imaging, diagnostics, and drug delivery applications due to their well-defined size, ease of functionalization, and high surface area. Quantum dots, in particular, offer excellent optical properties that make them useful for fluorescence-based imaging techniques.

One-Dimensional (1D) Nanomaterials One-dimensional nanomaterials, such as nanotubes and nanowires, possess a length that is significantly larger than their width and height. Carbon nanotubes (CNTs) are the most common 1D nanomaterials. They are known for their mechanical strength, conductivity, and unique interaction with biological systems. However, their toxicity profile remains a subject of concern, especially in terms of their potential to cause cellular damage or accumulation in organs. Functionalization of CNTs can help reduce their toxicity and enhance their biocompatibility.

Two-Dimensional (2D) Nanomaterials Two-dimensional nanomaterials, like graphene and graphene oxide, have a single layer of atoms or molecules, giving them extraordinary electrical, thermal, and mechanical properties. Their large surface area and flexibility make them ideal candidates for applications in drug delivery, biosensing, and tissue engineering. However, their biocompatibility and potential toxicity require careful consideration, and surface modification strategies are often employed to reduce cytotoxicity.

Three-Dimensional (3D) Nanomaterials Three-dimensional nanomaterials are characterized by their extended structure in all three spatial dimensions. These materials, such as nanoparticle aggregates and nanofibrous scaffolds, are often used in tissue engineering, as their complex structure allows for better cellular attachment, growth, and differentiation. Nanofibers, for example, can be engineered to replicate the structure of natural extracellular matrices, promoting tissue regeneration and wound healing.

3.3 Synthesis Methods of Biocompatible Nanomaterials

The synthesis of biocompatible nanomaterials involves various methods to ensure that the materials not only meet the required size, shape, and morphology but also exhibit desirable biological properties like low toxicity, high stability, and ease of functionalization. Nanomaterials can be synthesized via chemical, physical, or green/biological approaches. Each method offers specific advantages and challenges, depending on the application and desired outcomes. Furthermore, surface functionalization and post-synthesis modifications are crucial steps to enhance the biocompatibility and performance of the synthesized materials in biological systems.

3.3.1 Chemical Synthesis Methods

Chemical methods offer high control over particle size, morphology, and composition, making them widely used in the synthesis of nanomaterials. These techniques often involve the use of chemical reactions to reduce or deposit materials, providing excellent reproducibility and scalability [21–25].

3.3.1.1 Sol-Gel Process

The sol-gel process is one of the most widely used methods for synthesizing biocompatible nanomaterials, particularly ceramic and metal oxide nanoparticles. In this process, metal alkoxides (or other precursors) are hydrolyzed and polymerized to form a gel, which is then dried to produce the final nanomaterial. The sol-gel technique offers advantages such as low processing temperatures, excellent homogeneity, and the ability to control the size and composition of nanoparticles. It is frequently employed to synthesize silica, titania, and zirconia nanoparticles, which are used in drug delivery, imaging, and biosensing. However, one drawback of this method is the need for post-synthesis heat treatments to improve the material's crystallinity.

3.3.1.2 Co-Precipitation

Co-precipitation is a simple and cost-effective method for the synthesis of metal nanoparticles and metal oxide nanoparticles, often used to produce materials like iron oxide (Fe_3O_4) nanoparticles for biomedical applications. In this method, a solution containing metal salts is mixed with a precipitating agent (e.g., sodium hydroxide or ammonia) under controlled conditions, leading to the precipitation of nanoparticles. The advantage of co-precipitation lies in its simplicity and the possibility of scaling up for large-scale production. However, controlling particle size and morphology can be challenging, as these factors are influenced by the concentration of reactants and the reaction conditions.

3.3.1.3 Microemulsion

Microemulsion techniques are used to synthesize noble metal nanoparticles, such as gold (Au) and silver (Ag) nanoparticles, and oxide nanoparticles like zinc oxide (ZnO). In this method, nanoscale droplets of a water phase are dispersed in an oil phase using surfactants. This results in the formation of a stable emulsion that can serve as a reaction medium for nanoparticle synthesis. The advantage of microemulsion synthesis lies in the ability to control the size and distribution of nanoparticles with high precision. It is commonly used to synthesize materials that require strict size control for applications in drug delivery and imaging. However, the use of surfactants and solvents can pose challenges in ensuring the final product's biocompatibility and ease of surface modification.

3.3.2 Physical Synthesis Methods

Physical synthesis methods rely on physical processes, such as heat, light, or mechanical force, to produce nanoparticles. These techniques generally involve the application of energy to break down bulk materials into nanoscale components.

3.3.2.1 Laser Ablation

Laser ablation is a technique that uses a high-energy laser beam to vaporize a target material (typically a bulk metal or metal oxide) in a liquid medium. The vaporized atoms then condense to form nanoparticles. This method allows for precise control over particle size, making it suitable for synthesizing nanoparticles with specific properties required for applications in drug delivery, biosensing, and diagnostics. The major advantages of laser ablation include the ability to produce nanoparticles with high purity and without the need for chemical reagents. However, its scalability can be limited by the high cost of laser equipment, and the need for strict control over experimental conditions to prevent particle aggregation.

3.3.2.2 Mechanical Milling

Mechanical milling, also known as ball milling, is a widely used technique for producing nanoparticles by grinding bulk materials using mechanical force. In this method, a material is placed in a rotating container with milling media (balls), which collide with the material to break it down into nanoscale particles. Mechanical milling is particularly useful for producing metal oxide nanoparticles and other inorganic materials. The advantage of this method is its simplicity, scalability, and the ability to produce large quantities of nanoparticles. However, it often results in broad particle size distributions and may introduce defects or impurities that affect the material's biocompatibility.

3.3.3 Green/Biological Synthesis Methods

Green synthesis methods are considered environmentally friendly alternatives to conventional chemical and physical synthesis techniques. These methods often utilize biological agents such as plants, microorganisms, or enzymes to facilitate the formation of nanomaterials in a sustainable and nontoxic manner.

3.3.3.1 Plant-Mediated Synthesis

Plant-mediated synthesis of nanomaterials involves using plant extracts to reduce metal ions and form nanoparticles. This method is particularly attractive due to its eco-friendly nature and the availability of plant-based reducing agents that can lead to the formation of highly biocompatible nanoparticles. Common examples include the use of tea extracts, citrus peel, or plant leaves for the synthesis of gold, silver, and copper nanoparticles. Plant-based synthesis offers several advantages, including simplicity, low cost, and the ability to control particle size. Additionally, plant extracts often contain bioactive compounds that enhance the biocompatibility of the nanoparticles. However, standardization of the process and the identification of active compounds in plant extracts remain challenges in the field.

3.3.3.2 Microbial Synthesis

Microbial synthesis involves the use of microorganisms such as bacteria, fungi, and algae to produce nanoparticles. These organisms can facilitate the reduction of metal ions to form nanoparticles, often under mild conditions. For instance, bacteria like *Pseudomonas* and *Escherichia coli* have been employed to produce silver and gold nanoparticles. Fungal species, such as *Aspergillus* and *Penicillium*, are also widely used for synthesizing nanoparticles with unique properties. The main advantages of microbial synthesis include high reproducibility, scalability, and the ability to produce nanoparticles with well-defined characteristics. Additionally, the biocompatibility of these nanoparticles is enhanced by the natural processes occurring within the microorganisms. However, challenges include the need for proper control of microbial growth conditions to ensure the consistency of nanoparticle production.

3.3.3.3 Enzyme-Assisted Synthesis

Enzyme-assisted synthesis involves the use of enzymes as catalysts to facilitate the reduction of metal ions and the formation of nanoparticles. This method offers high precision and control over the synthesis process, as enzymes can target specific chemical reactions. Enzyme-assisted synthesis is particularly useful for producing nanoparticles with enhanced biocatalytic activity and for applications where bioactive materials are required. The main advantage of this method is its mild reaction conditions and ability to produce well-defined nanoparticles. However, the enzyme costs and potential difficulties in enzyme recovery and reuse can limit the practicality of this method for large-scale applications.

3.3.4 Surface Functionalization and Post-Synthesis Modification

Surface functionalization refers to the modification of the surface of nanomaterials to enhance their biocompatibility, stability, and targeting ability. After the synthesis of nanomaterials, their surfaces often need to be modified to improve their interaction with biological systems [26–28].

Polymer coating: One common approach is to coat nanoparticles with biocompatible polymers such as polyethylene glycol (PEG), polyvinyl alcohol (PVA), or chitosan. These coatings help stabilize the nanoparticles and reduce nonspecific interactions with biological molecules, thus preventing toxicity and enhancing circulation time in the bloodstream.

Targeting ligands: Surface modification can also include the conjugation of targeting molecules (e.g., antibodies, peptides, or aptamers) to nanoparticles. This allows for the specific targeting of disease sites or cells, such as cancer cells, enhancing the therapeutic efficacy of the nanomaterial.

Metallic and organic functional groups: The incorporation of functional groups like amine, carboxyl, or thiol groups on the nanoparticle surface can also allow for further conjugation with biomolecules, drugs, or diagnostic agents, facilitating their use in a range of biomedical applications.

3.4 Challenges and Limitations

The use of biocompatible nanomaterials in biomedical and environmental applications is hindered by several challenges. Toxicity and safety remain major concerns due to the unique properties of nanomaterials, such as their small size and large surface area, which may cause unintended biological interactions, toxicity, or environmental risks. Additionally, achieving reproducibility and scalability in the synthesis of nanomaterials is difficult, as small changes in reaction conditions can lead to inconsistent material properties, and scaling up from lab to industrial levels often faces issues in maintaining quality and cost-effectiveness. Regulatory and ethical challenges also pose obstacles, as existing frameworks often do not address the specific characteristics of nanomaterials, creating uncertainty in approval processes. Ethical issues, such as potential misuse or environmental impacts, add complexity to the adoption of these materials. Overcoming these challenges requires continued research, the establishment of standardized regulations, and public engagement to ensure the safe and effective application of nanomaterials.

3.5 Conclusion

In conclusion, biocompatible nanomaterials offer significant promise for advancing biomedical and environmental applications due to their unique characteristics, such as high surface area, customizable size, and ease of surface modification. However, challenges related to toxicity, reproducibility, scalability, and regulatory uncertainties need to be overcome to fully realize their potential. To ensure these materials are both safe and effective for practical use, comprehensive toxicological evaluations, standardized synthesis methods, and robust regulatory frameworks must be developed. Additionally, ethical considerations regarding their long-term effects and societal impact must be factored into their development and deployment. With continued research, innovation, and responsible regulation, biocompatible nanomaterials can be successfully integrated into a wide range of applications, benefiting human health and the environment.

References

1. Al-Radadi NS (2021) Facile one-step green synthesis of gold nanoparticles (AuNp) using licorice root extract: antimicrobial and anticancer study against HepG2 cell line. Arab J Chem 14(2):102956. https://doi.org/10.1016/j.arabjc.2020.102956
2. Vadlapudi V, Behara M, Nagalakshmi Devamma M (2014) Green synthesis and biocompatibility of nanoparticles. Rasayan J Chem 7:219–223
3. Al-Radadi NS (2021) Green biosynthesis of flaxseed gold nanoparticles (au-NPs)as potent anti-cancer agent against breast cancer cells. J Saudi Chem Soc 25(6):101243. https://doi.org/10.1016/j.jscs.2021.101243
4. Al-Radadi NS (2022) Microwave assisted green synthesis of Fe@au core-shell NPs magnetic to enhance olive oil efficiency on eradication of helicobacter pylori (life preserver). Arab J Chem 15(5):103685. https://doi.org/10.1016/j.arabjc.2022.103685
5. Shah KW, Huseien GF (2020) Inorganic nanomaterials for fighting surface and airborne pathogens and viruses. Nano Express 1(3):032003. https://doi.org/10.1088/2632-959X/abc706
6. Al-Radadi NS (2022) Laboratory scale medicinal plants mediated green synthesis of biocompatible nanomaterials and their versatile biomedical applications. Saudi J Biol Sci 29(5):3848–3870. https://doi.org/10.1016/j.sjbs.2022.02.042
7. Salem SS, Hammad EN, Mohamed AA, El-Dougdoug W (2023) A comprehensive review of nanomaterials: types, synthesis, characterization, and applications. Biointerf Res Appl Chem 13(1):41. https://doi.org/10.33263/BRIAC131.041
8. Jana S, Bandyopadhyay A, Datta S, Bhattacharya D, Debnarayan J (2021) Emerging properties of carbon based 2D material beyond graphene. J Phys Condens Matter. https://doi.org/10.1088/1361-648X/ac3075
9. Mohamed AE-MA, Mohamed MA (2020) Carbon nanotubes: synthesis, characterization, and applications. In: Carbon nanomaterials for agri-food and environmental applications. Elsevier, Amsterdam, pp 21–32. https://doi.org/10.1016/B978-0-12-819786-8.00002-5
10. Foong LK, Foroughi MM, Mirhosseini AF, Safaei M, Jahani S, Mostafavi M, Ebrahimpoor N, Sharifi M, Varma RS, Khatami M (2020) Applications of nano-materials in diverse dentistry regimes. RSC Adv 10:15430–15460. https://doi.org/10.1039/D0RA00762E

11. Ijaz I, Gilani E, Nazir A, Bukhari A (2020) Detail review on chemical, physical and green synthesis, classification, characterizations and applications of nanoparticles. Green Chem Lett Rev 13:223–245. https://doi.org/10.1080/17518253.2020.1802517
12. Singh V, Yadav P, Mishra V (2020) Recent advances on classification, properties, synthesis, and characterization of nanomaterials. Green Synth Nanomater Bioenergy Appl:83–97. https://doi.org/10.1002/9781119576785.ch3
13. Kankala RK, Han YH, Na J, Lee CH, Sun Z, Wang SB, Kimura T, Ok YS, Yamauchi Y, Chen AZ (2020) Nanoarchitectured structure and surface biofunctionality of mesoporous silica nanoparticles. Adv Mater 32:1907035. https://doi.org/10.1002/adma.201907035
14. Fouda A, El-Din Hassan S, Salem SS, Shaheen TI (2018) In-vitro cytotoxicity, antibacterial, and UV protection properties of the biosynthesized zinc oxide nanoparticles for medical textile applications. Microb Pathog 125:252–261. https://doi.org/10.1016/j.micpath.2018.09.030
15. Al-Hakkani MF (2020) Biogenic copper nanoparticles and their applications: a review. SN Appl Sci 2:1–20. https://doi.org/10.1007/s42452-020-2279-1
16. Wang F, Guan W, Xu L, Ding Z, Ma H, Ma A, Terry N (2019) Effects of nanoparticles on algae: adsorption, distribution, ecotoxicity and fate. Appl Sci 9:1534. https://doi.org/10.3390/app9081534
17. Hou Z, Liu Y, Xu J, Zhu J (2020) Surface engineering of magnetic iron oxide nanoparticles by polymer grafting: synthesis progress and biomedical applications. Nanoscale 12:14957–14975. https://doi.org/10.1039/D0NR03346D
18. Williams DF (2008) On the mechanisms of biocompatibility. Biomaterials 29(20):2941–2953. https://doi.org/10.1016/j.biomaterials.2008.04.023
19. Iravani S (2011) Green synthesis of metal nanoparticles using plants. Green Chem 13(10):2638–2650. https://doi.org/10.1039/C1GC15386B
20. Buzea C, Pacheco II, Robbie K (2007) Nanomaterials and nanoparticles: sources and toxicity. Biointerphases 2(4):7–71. https://doi.org/10.1116/1.2815690
21. Ayode Otitoju T, Ugochukwu Okoye P, Chen G, Li Y, Onyeka Okoye M, Li S (2020) Advanced ceramic components: materials, fabrication, and applications. J Ind Eng Chem 85:34–65. https://doi.org/10.1016/j.jiec.2020.02.002
22. Dacrory S, Hashem AH, Hasanin M (2021) Synthesis of cellulose based amino acid functionalized nanobiocomplex: characterization, antifungal activity, molecular docking and hemocompatibility. Environ Nanotechnol Monit Manag 15:100453. https://doi.org/10.1016/j.enmm.2021.100453
23. Noore S, Rastogi NK, O'Donnell C, Tiwari B (2021) Novel Bioactive Extract Nano-Encapsul Encycl 1:632–664. https://doi.org/10.3390/encyclopedia1030052
24. Zhuang S, Lee ES, Lei L, Nunna BB, Kuang L, Zhang W (2016) Synthesis of nitrogen-doped graphene catalyst by high-energy wet ball milling for electrochemical systems. Int J Energy Res 40:2136–2149. https://doi.org/10.1002/er.3595
25. Li X, Wang L, Fan Y, Feng Q, Cui F-z (2012) Biocompatibility and toxicity of nanoparticles and nanotubes. J Nanomater 2012(1):548389. https://doi.org/10.1155/2012/548389
26. Selvakesavan RK, Franklin G (2021) Prospective application of nanoparticles green synthesized using medicinal plant extracts as novel nanomedicines. Nanotechnol Sci Appl:179–195. https://doi.org/10.2147/NSA.S333467
27. Annamalai J, Ummalyma SB, Pandey A, Bhaskar T (2021) Recent trends in microbial nanoparticle synthesis and potential application in environmental technology: a comprehensive review. Environ Sci Pollut Res 28(36):49362–49382. https://doi.org/10.1007/s11356-021-15680-x
28. Sutradhar S, Mondal S, Ghosh BN (2023) Available synthesis methods of green nanomaterials, their properties, and characterization. In: Nanomaterials: the building blocks of modern technology: synthesis, properties and applications. Springer, Singapore, pp 211–229. https://doi.org/10.1007/978-981-99-4149-0_11

Chapter 4
Characterizations Techniques of Biocompatible Nanomaterials

Arunkumar Lagashetty, H. Manjushree, B. Amrutha Bindu, K. Diwakar, P. V. M. Goutham, and B. M. Roja

Abstract Nanomaterials have attracted much to the recent researcher's interest as a rapidly growing class of materials for many applications. Especially biologically derived nanomaterials construct many application-oriented properties by simple way of fabrication. Characterization of such materials for their phase is considered to be an important aspect to define properties and applications of materials. Many techniques have been used to characterize the materials for their confirmation, physical properties, and nature of biocompatible nanomaterials. In several cases, more than one characterization technique are supposed to be used to evaluate the materials for their identification. It is a well-known fact that the fabrication of bioderived nanoparticles are constantly increasing, hence it is a challenging to the researchers for reliable characterization of nanomaterials for specialized applications. The aim of this chapter is to summarize the importance of experimental techniques for the characterization of biocompatible nanoparticles. Principles and instrumentation of some important characterization techniques are summarized. Also, the main characteristics of the techniques and their operation principles in characterization of biomaterials are described.

Keywords Nanomaterials · Characterization · Bioderived · Biocompatible · Biomaterials

A. Lagashetty (✉) · H. Manjushree · B. Amrutha Bindu · K. Diwakar · P. V. M. Goutham
Department of Chemistry, Vijayanagara Sri Krishnadevaraya University, Ballari, Karnataka, India

B. M. Roja
Department of Physics, Vijayanagara Sri Krishnadevaraya University, Ballari, Karnataka, India

R. Naik et al. (eds.), *Biocompatible Nanomaterials for Smart Sensors*, Nanostructure Science and Technology,
https://doi.org/10.1007/978-3-032-01745-1_4

4.1 Introduction

Nanosized biocompatible materials are a class of materials that can be safely used in contact with living tissues without causing any harm to animal body [1, 2]. These materials are synthesized by simple bio reduction technique like green synthesis to minimize environmental toxicity [3, 4]. Green derived materials have many applications including environmental, food safety, drug delivery, and tissue engineering to solve societal problems [5–8]. They can be used to enhance biological reaction systems like food quality, improvement in drug delivery systems, and plant biotechnology [9]. A major activity of biocompatible nanomaterials is their ability to interact with living tissues without causing adverse effects. These materials are designed to be nontoxic, nonirritating, and biodegradable. Biocompatible nanomaterials are produced by green synthesis method using natural sources like bacteria or plants. The environmental impact of these techniques is becoming more and more significant [10, 11]. Nanomaterials derived by natural biological process are considered as bioproducts which are more in number compared to chemically derived nanomaterials. Nanostructures, which are comparable in the size to biological molecules, have been recognized since the past ten decades [12, 13]. Innovations in nanoscience routes more discussion on fundamental scientific phenomena and nature of molecules less than 100 nm. Nanosized biocompatible materials have several benefits compared to traditional materials. Small size and high surface area properties improve many properties followed by their applications. These materials can be employed as contrast agents for imaging, enabling the real-time observation of molecular biological processes. Applications of biocompatible nanomaterials in medicine depend on numerous characterization techniques which are discussed.

Characterization techniques are methods used to identify materials or to characterize materials using microscopy, spectroscopy, and other instrumental measurements [14]. Many researchers have reviewed that characterization tools are considered as important measurement methods to achieve the materials confirmation [15, 16]. These are used to understand the phase formation and properties of materials at the nano scale. Characterization techniques like microscopy and spectroscopy are used to analyze a wide range of nanomaterials [17]. Future research problems in materials confirmation using different instruments are crucial for the worldwide community. Advanced researchers need modern characterization techniques for differently fabricated nanomaterial analysis to optimize the new procedures. The importance of these analyzing methods can be related to properties being related to their structure. Various materials at varied shapes and sizes can be characterized by modern instrumental techniques [18, 19]. Varied inputs for the tools based on the nature of the materials under characterization prove the facts and steps for their utilization. Preparation or fabrication of materials is an important step in exploration of materials information in the form of desired results. In addition, other factors like scan rate, time, base materials, etc., can also influence the obtained results. In the sections that follow, the various discuss the different characterization tools like X-ray diffraction (XRD), scanning electron microscopy (SEM), Fourier transform -infrared (FT-IR), Raman spectroscopy (RS), transmission electron microscopy (TEM), UV-visible spectroscopy, energy dispersive x-ray microanalysis (EDX) for the biocompatible materials are elaborated systematically.

4.2 X-ray Diffraction

Interaction of X-rays with nanomaterials forms diffraction called materials diffraction, whose results provide much information about materials. The diffraction pattern depends on the nature of the material and its structural arrangement. Crystal diffraction works on the basis of Bragg's law: 2dsin $\theta = n\lambda$, where n is an integer, λ is the wavelength of the radiation, d is the spacing between surfaces, and θ is the angle of diffraction. This relation signifies the only way to observe interference effects is when radiation interacts with physical dimensions. Any crystalline material can be identified and analyzed with the use of X-ray spectra. The quality of the output will be reflected in the degree of crystallinity. Generally, X-ray diffractometer is made up of X-ray generator, a goniometer, sample holder, and an X-ray detector. Tool emitted X-rays are produced by X-ray tubes, which strike a metal object with high-energy electrons that eliminate core electrons. This technique usually analyzes the polycrystalline materials and powdery materials. According to Bragg's rule, the evolution of reflections in the experiment that correspond to material planes. Powder diffractometers can be used to view the results of diffraction of flour patterns. In addition, a powder camera method is simpler and can provide more information than a powder diffractometer. In order to identify the phase of the material, the obtained spectrum is systematically compared to the most recent standard issued by the Joint Committee of Powder Diffraction Standard. This analyzing system is used to study the structure of biologically derived samples. Metal oxides, metals, and metal oxide nanocomposites can be studied better by bombarding X-rays on the samples. Bragg's reflection in the form of a pattern gives the information about the nature of the sample. It is also a crucial technique for analyzing biocompatible nanomaterials, providing insights into their structural properties, including crystallinity, phase, and grain size. The structure of biomaterial composites that contain embedded nanoparticles, providing insights into the interaction between the biomaterial and the nanoparticles were well studied by many researchers. Further, it can be used to characterize the crystalline structure of nano medicines, which can be important for understanding their stability and release behavior. Also, in situ performance of the tool one can monitor the changes in the structure of nanomaterials during biological process. We attempted the synthesis of copper nanoparticles (Cu NPs) by bioreduction reaction using green route. *Pandanus tectorius* leaf extract is used to prepare CuNPs from its salt solution. Figure 4.1 shows indexed XRD pattern of green-derived Cu NPs. The material's crystalline nature is demonstrated by the Bragg's reflection pattern. Obtained 2*d* values of the sample

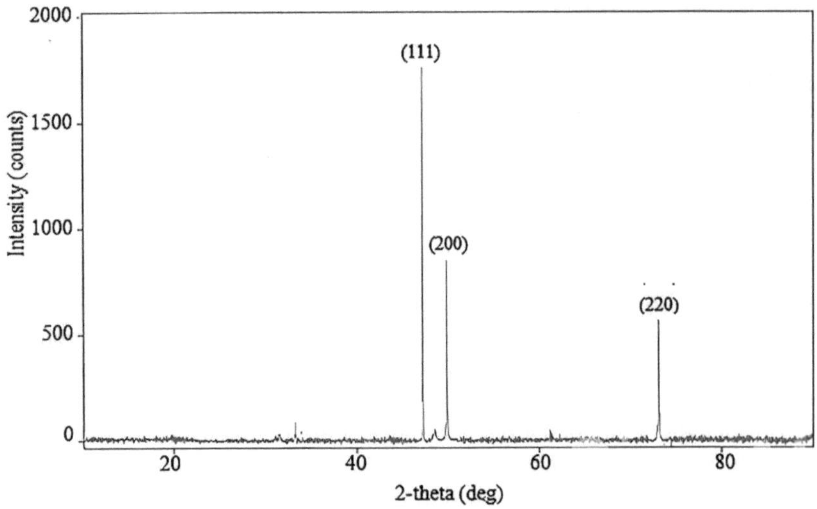

Fig. 4.1 XRD pattern of Cu NPs

are matched well with JCPDS file 04–0836 of Cu NPs. Hence the pattern confirms the formation of green-derived Cu NPs.

4.3 Fourier Transform Infrared

Fourier-transform infrared spectroscopy (FT-IR) is a technique used to obtain an infrared spectrum of absorption or emission of a solid, liquid, or gas. The bonding nature of materials may be characterized with this spectroscopic approach, which is particularly helpful. It provides broad application, vibrational measurements in the ambient environment, and comprehensive structural information. Absorption spectra are created when two vibrational levels in the ground state shift, giving rise to a spectrum. Two broad regions—the functional group area and the fingerprint region—can be used to effectively study the infrared spectrum. The region of the first functional group is typically treated as characteristic of the fingerprint region. Molecular vibrations, usually bending motions that are characteristics of the entire molecule or large fragments of the molecule can be observed in fingerprint region. The functional group region tends to include stretching vibrations, which are more localized and characteristic of the typical functional groups. While these bands are not very useful in confirming identity, they do provide some very useful information about the nature of the components that make up the molecule. Usually, an IR spectrometer consist of major three parts, viz., source, monochromator, and the receptor. Between the source and the detector there must be some kind of device to analyze the radiation so that intensity can be evaluated for each wavelength resolution element. A Fourier-Transform Infrared (FTIR) spectrometer concurrently gathers data with excellent spectral precision across a broad spectrum. This offers a major benefit over a dispersive spectrometer, which gauges intensity throughout a small wavelength range at once. IR spectral data of prepared materials were recorded at room temperature for wave number of 4000 to 400 cm^{-1} spectrum obtained by using KBR press. This characterization tool finds its importance in identification of biocompatible materials like green-derived materials. This spectroscopic technique is a valuable technique for characterizing biocompatible nanomaterials. It can be achieved by analyzing molecular vibrations and functional groups, providing insights into their chemical composition and interactions. Here is an attempt to characterize the bioderived Cu NPs using green route. FT-IR spectra sample was conducted in the 4000–5000 cm^{-1} spectral area, as can be seen in Fig. 4.2 individually. Many important absorption peaks may be seen in the spectrum as a result of different vibrational bonds. Typically, the solid sample's bands are attributed to ion vibration inside the crystal lattice. The copper sample's wide band at around 3421 cm^{-1} is ascribed to the stretching modes and H–O–H vibration of the water that was absorbed. The C=O stretching vibrations are also assigned a frequency of 1630 cm^{-1}. About 626 cm^{-1} of absorption is associated with stretching vibrations between metals. The

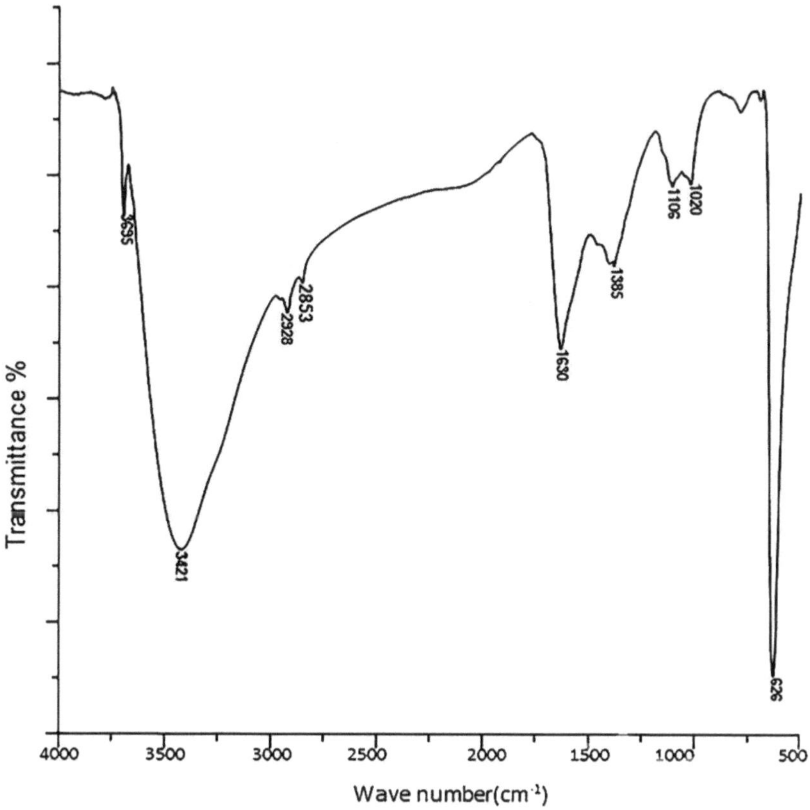

Fig. 4.2 FT-IR spectrum of Cu NPs

powdered bimetallic oxide is a nanocrystal based on the width of the absorption band. C–O and C–H bending are shown by bands at 1385 cm^{-1} and 1106 cm^{-1}, respectively.

4.4 Scanning Electron Micrograph

A scanning electron microscope (SEM) is an electron microscope used to study the surface morphology of the sample. Morphological mages can be obtained by a sample in which it is focused by electron beams. The sample interacts with the electronic beam and produces different information signals about the surface morphology and composition. Electronic scan pattern with the intensity of the detected signal is used to produce an image. It may be possible that secondary electrons emitted by sample atoms excited by the electron beam are detected using a secondary electron detector. The topography of the specimen determines the signal intensity led by detected secondary electrons. Some scanning tools can achieve high-resolution images at nanoscale. The said microscopic tool is a scanning technique used to analyze biologically derived nanomaterials. It provides high-resolution images to visualize surface characteristics of biocompatibility materials. Here is an attempt to analyze bioproducts using SEM for its surface morphology. Figure 4.3 shows SEM image of green-derived Cu NPs. The image shows irregular shaped particles

Fig. 4.3 SEM image of Cu NPs

forming a close streak-like arrangement, the particles exhibit a range of sizes and compact, closely-assembled shapes.

4.5 Transmission Electron Microscopy

It is also an electronic tool that produces enlarged images at smaller wave length for high resolution. In the field of materials science and technology, transmission electron microscope stands out as the most used characterization method for immediate imaging of solid samples. It is a widespread characterization device for analyzing the morphological structure of the samples at nanoscale. The TEM tool's chosen area electron diffraction (SAED) pattern type validates the crystalline plane. The highest point of the electron gun of the transmission electron microscope radiates high intense beam of electrons in to the vacuum in which it accelerates between the cathode and the anode. After that it goes through a segment of the magnifying instrument under high vacuum condition. Transmission electron microscope tool consists of three major components such as electron gun, image producing system, and imaging recording system. Electron gun at the top of a TEM tool emits electron beam that penetrates through vacuum tube. The tool consists of objective lens, a movable holder, and projector lenses. They function by focusing the passing electron through the specimen forming a highly magnified image. The image recording system consists of florescent screen used to view and to focus on the image. It is fitted with digital camera that can effectively capture the image. They have vacuum system that prevents the bombardment or collision of electrons with air molecules disrupting their movement and ability to focus. The observed image is considered as monochromatic image. An electromagnetic lens to point the electrons into a thin beam is used in the TEM tool. This electron beam penetrates through the analyzing sample, which is dispersed in the carbon covered copper grid. Contingent upon the thickness of the material present, few of the electrons are scattered and vanish from the beam. At the base of the microscopic instrument, an unscattered electron hit a fluorescent screen, which offers ascents to an image of the sample showed with different obscurity, as per their thickness. The image of the particle can be recorded as photo or computerized image for further analysis. To examine the sample, SAED pattern is used for crystal structure and crystal defects of the sample. It is a valuable tool for analyzing biocompatible nanomaterials by particle morphology, size, and structure. TEM allows for direct imaging of nanoparticles and their interactions with cellular and tissue components, providing detailed information crucial for biocompatibility studies. It is very informative tool used to analyze biocompatible nanomaterials, offering insights into their particle morphology at the nanoscale. This microscopy allows for direct imaging of nanoparticles and their interactions with cellular and tissue components, providing detailed information crucial for biocompatibility studies. This chapter attempts to analyze bioproducts using TEM for their particle morphology. Figure 4.4 shows TEM image of green-derived Cu NPs.

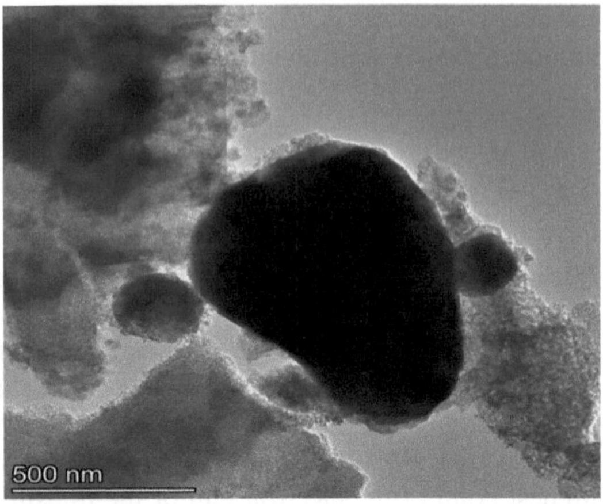

Fig. 4.4 TEM image of Cu NPs

Most of the particles are in spherical shape and also represent the crystalline nature of the sample that falls under nanoscale.

4.6 Raman Spectroscopy

Raman spectroscopy is a spectroscopic technique typically used to determine vibrational mode, rotational mode, and other low frequency modes of systems. It is commonly used in chemistry to provide a structural fingerprint by which molecules can be analyzed and identified. Raman spectroscopy depends on Raman scattering, which means inelastic scattering of photons. Monochromatic laser light sources interact with sample resulting in the energy of the laser photons being shifted. The shift in energy gives information about the vibrational modes in the system. When the molecule is placed in an electromagnetic field, a transfer of energy from the field to the molecule will occur only when the difference in energy between two quantified states equals to Planck's constant and the frequency of light. Raman microscopes are now widespread because they offer nondestructive and quantitative microanalysis of structures and electronic properties of various samples. Raman spectra originate in the electronic polarization caused by monochromatic light source. The incident radiation interacts with the molecular vibrational energy levels has its origins in light scattering, in which the electromagnetic radiation interacts with a polarizable electron cloud. This interaction modulated by the molecular vibrations leads to scattering. Raman spectrometer consists of illumination source composed of laser emitting source, a macrosample chamber, and a classical optical microscope for microanalysis. Double monochromator and a spectrograph is also composed to the tool, which can be used either in subtractive mode to provide a wide field to the spectrograph or in additive mode when higher resolution is required. Further, the detection system includes a multichannel detector and a photomultiplier connected to a single photon counter. Raman spectroscopy is a very important

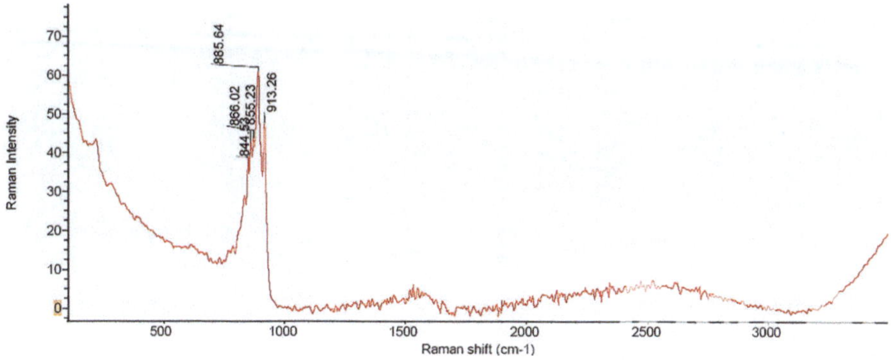

Fig. 4.5 Raman spectrums of Cu NPs

instrument used for analyzing biocompatible nanomaterials for structure, composition, and properties. It is also used to identify crystalline phases and defects nanomaterials. Also, it is making possible to detect nanomaterials and their interactions with biological systems. Raman spectroscopy is an important technique which can be used to study chemical identification, characterization of molecular structures, and effects of bonding environment of the sample. Here is an attempt to analyze the biocompatible product using Raman spectroscopic tool. Figure 4.5 shows Raman spectrum of Cu NPs prepared by *Pandanus tectorius* plant leaf extract. Figure shows two intense peaks at 855.23 and 913.26 that are evident, and these can be ascribed to the Cu–O vibrations for sub-surface species and O–O mode for adsorbed molecular oxygen. Molecular oxygen adsorption may be facilitated by subsurface species, although this could also be a likely effect of surface restriction in Cu NPs.

4.7 Energy Dispersive X-ray Spectroscopy (EDAX)

An effective analytical technique for the elemental analysis of the sample is energy-dispersive X-ray spectroscopy, commonly referred to as dispersive X-ray analysis (EDXA or EDAX) or energy dispersive X-ray microanalysis (EDXMA). It is based on the interaction between the sample materials and X-ray excitation. It operates on the core tenet that every element has a distinct atomic structure that permits a unique set of peaks on its electromagnetic emission spectrum. The basic principle behind this instrumentation is the interaction between electrons with the analyzing atoms of the material. Electrons bombarded with material leads inner-shell electrons of the sample to be excited and return to their original position by releasing energy. The energy of these emitted radiation characteristic of the elements present in the sample material. Instrumentation consists of a detector fixed on the electron microscope usually made up of solid-state silicon drift detector. It finds its importance in determination of the energy and intensity of the emitted X-rays, when bombarded on the

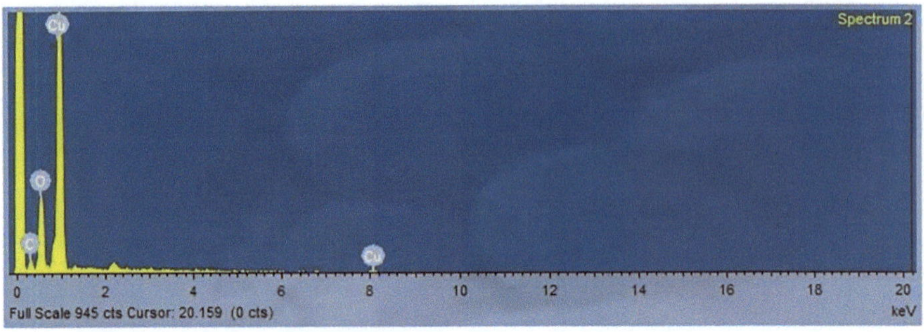

Fig. 4.6 EDX pattern of CuNPS

sample. Observed peaks in the spectrum are due to emitted X-rays by different elements in the analyzing sample materials. Standard data base is used to identify elemental peaks observed in the spectrum. Qualitative and quantitative signals of the elements present in the sample can also be observed. The intensity of the X-ray peaks can be measured, and by comparing with standards it helps to analyze the elements in sample. EDAX is a valuable tool in the analysis of biocompatible nanomaterials, allowing for the determination of elemental composition. It can identify elements present in the nanomaterial and their relative quantities, providing insights into the material's structure and potential biocompatibility. Here is an attempt to characterize the biocompatible CuNPS derived by green route. To determine whether Cu NPs are present in the sample, the EDX technique has been utilized for conformational analysis. Figure 4.6 shows EDX pattern of green-derived Cu NPs sample. Signals that came from the grid are used for the sample. The Cu atom signal is present at the appropriate binding energy in the figure. The pattern's absorption peak with these features also validates the creation of the Cu NPs sample.

4.8 UV-Visible Spectroscopy

Ultraviolet-visible (UV-Vis) spectroscopy is a technique for measuring the amount of visible and ultraviolet light that an analyzing sample absorbs. It detects different chemicals' existence and concentration when they dissolve in a solvent. The working principle of this spectroscopy is the interaction of light radiations with analyzing molecules. The sample is made in contact with UV or visible light; electrons in the molecules can absorb the energy from the light and exited to states of higher energy. The energy of the absorbed light, which is in the visible or ultraviolet spectrum, is equal to the difference in energy between the ground and excited states. The main components of the UV-Vis spectrometer tools are light source, sample holder, monochromatic light system, and detector. Usually light source is made up of a deuterium lamp as a UV source and a tungsten lamp as visible light source. A transparent cuvette made of quartz or glass is used as a sample holder, which allows light

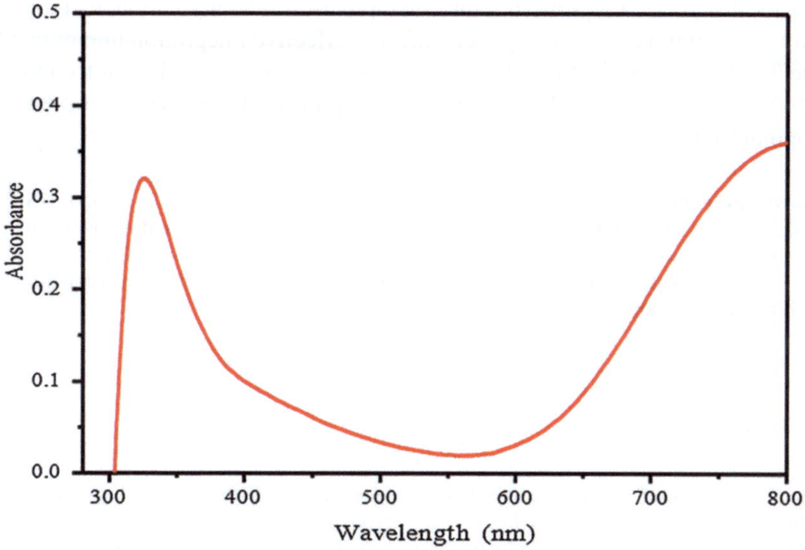

Fig. 4.7 UV spectrum of Cu NPs

to pass through it. The wavelength of light is specified by a monochromatic system tailored to the tool. It separates the white light into individual according to the desired wavelength. Intensity of the absorbed light is measured by the detector. Data obtained from the detector are processed and displayed using specialized software, which provides the UV-Vis spectrum of the sample. This spectrophotometric technique is used in many disciplines of science to analyze the samples. Also, it is used to get an idea about the formation of nanomaterials by its absorption. Figure 4.7 shows UV-visible spectrum of Cu NPs to view and confirm the formation of particles in the initial stage. It showed one absorbance bands at 330 nm. The good development of this band indicates that the Cu NPs were successfully formed by green route.

4.9 Conclusion

Biocompatible nanomaterials, particularly those derived from biological sources, hold immense promise for a wide range of applications due to their unique physicochemical properties and ease of fabrication. However, the accurate characterization of these materials is essential to determine their structure, composition, and functional capabilities, which directly influence their performance in specialized applications. This chapter has reviewed key experimental techniques used for characterizing biocompatible nanomaterials, highlighting their principles, instrumentation, and suitability for different material types. As the development of such materials continues to grow, reliable and precise characterization methods will remain a cornerstone in ensuring their safe and effective integration into biomedical and technological applications. Continued advancements in analytical technologies will further enhance our ability to explore and optimize the properties of these innovative materials.

Acknowledgments The authors convey their heartfelt thanks to DST-FIST (SR/FST/CSI-003/2016) for the grant provided to procure the required instruments and infrastructure at the Chemistry Department of Vijayanagar Sri Krishnadevaraya University Ballari. Our thanks are also due to Prof. A. Venkataraman, Department of Chemistry, Gulbarga University, Kalaburagi for useful direction and support.

References

1. Hamdi E, Hidouri S (2024) Biological uses of nanomaterials within the safe handling and toxic effects: (brain as a model). Adv Redox Res 12:100105
2. Saikia N (2024) Inorganic-based nanoparticles and biomaterials as biocompatible scaffolds for regenerative medicine and tissue engineering: current advances and trends of development. Inorganics 12:292
3. Kazemi S, Hosseingholian A, Gohari SD, Feirahi F, Moammeri F, Mesbahian G, Moghaddam ZS, Ren Q (2023) Recent advances in green synthesized nanoparticles: from production to application. Mater Today Sustain 24:100500
4. Aarthye P, Sureshkumar M (2021) Green synthesis of nanomaterials: an overview. Mater Today Proc 47(4):907–913
5. Kaur M, Gautam A, Guleria P, Singh K, Kumar V (2022) Green synthesis of metal nanoparticles and their environmental applications. Curr Opin Environ Sci Health 29:100390
6. Egwu CO, Aloke C, Onwe KT, Umoke CI, Nwafor J, Eyo RA, Chukwu JA, Ufebe GO, Ladokun J, Audu DT (2024) Nanomaterials in drug delivery: strengths and opportunities in medicine. Molecules 29:2584
7. Zheng X, Zhang P, Fu Z, Meng S, Dai L, Yang H (2021) Applications of nanomaterials in tissue engineering. RSC Adv 11:19041–19058
8. Britto CGJ, John ADSA, Jayakumar A, Selvasekaran P, Chidambaram R, Zheng H (2023) Nanotechnology applications for food safety: benefits and risks. In: Sharma A, Vijayakumar PS, Pramod E, Prabhakar K, Kumar R (eds) Nanotechnology applications for food safety and quality monitoring. Academic, New York, pp 3–30
9. Holghoomi R, Colagar AH (2024) Applications of biocompatible nanoparticles in plant biotechnology for enhanced secondary metabolite biosynthesis. Inorg Chem Commun 167:112753
10. Zango ZU, Garba A, Shittu FB, Imam SS, Haruna A, Zango MU (2025) A state-of-the-art review on green synthesis and modifications of ZnO nanoparticles for organic pollutants decomposition and CO_2 conversion. J Hazard Mater Adv 17:100588
11. Lagashetty A, Ganiger SK, Preeti RK, Reddy S, Pari M (2020) Microwave-assisted green synthesis, characterization and adsorption study of metal oxide nanoparticles by ficus benghalensis plant leaf extract. New J Chem 14:14095
12. Alsaiari NS, Alzahrani FM, Amari A, Osman H, Harharah HN, Elboughdiri N, Tahoon MA (2023) Plant and microbial approaches as green methods for the synthesis of nanomaterials: synthesis, applications, and future perspectives. Molecules 28:463
13. Haripriyaa M, Suthindhiran K (2023) Pharmacokinetics of nanoparticles: current knowledge, future directions and its implications in drug delivery. Futur J Pharm Sci 9:113

14. Dutt A, Saini N, Kalia A, Madan P, Srikanth T, Talukdar S (2024) Biocompatible nanomaterials for sustainable biomedical applications. E3S Web Conf 547:03020
15. Mourdikoudis S, Pallares RM, Thanh NTK (2018) Characterization techniques for nanoparticles: comparison and complementarity upon studying nanoparticle properties. Nanoscale 10:12871–12934
16. Titus D, Samuel EJJ, Roopan SM (2019) Nanoparticle characterization techniques. In: Shukla AK, Iravani S (eds) Micro and nano technologies, green synthesis, characterization and applications of nanoparticles. Elsevier, Amsterdam, pp 303–319
17. Joudeh N, Linke D (2022) Nanoparticle classification, physicochemical properties, characterization, and applications: a comprehensive review for biologists. J Nanobiotechnol 20:262
18. Chaudhary J, Tailor G, Shailesh SK, Kumar D (2016) Characterization techniques used in nanomaterials - a review. J Environ Nanotechnol 5(2):30–35
19. Harbi A, Abd-Elrahman NK (2024) Physical methods for preparation of nanomaterials, their characterization and applications: a review. J Umm Al-Qura Univ Appl Sci 11:356–377

Chapter 5
Green Synthesized Nanocomposites for Mercury, Lead, and Cadmium Detection

G. S. Rakshitha, D. N. Varun, B. M. Santhosh, G. M. Bhavana, B. P. Sanjay, M. S. Kiran, P. Yuvarajagouda, and N. Hareesha

Abstract Green-synthesized nanocomposites represent a promising class of materials for the sensitive and selective detection of toxic heavy metals such as mercury (Hg^{2+}), lead (Pb^{2+}), and cadmium (Cd^{2+}). Derived from renewable biological resources including plant extracts and biopolymers, these nanomaterials offer a sustainable alternative to conventional sensors, combining environmental safety with high-performance functionality. This chapter highlights their synthesis routes, key structural and surface properties, and applications in electrochemical, colorimetric, and fluorescence-based sensing platforms. Case studies across water, soil, and biological matrices demonstrate their real-world applicability. While challenges in reproducibility and long-term stability persist, green nanocomposites hold significant potential for next-generation portable and eco-friendly sensor technologies.

G. S. Rakshitha
Department of Chemistry, SJCE, JSS Science and Technology University, Mysuru, Karnataka, India

D. N. Varun
Department of Civil, Environmental, and Mechanical Engineering, University of Trento, Trento, Italy

B. M. Santhosh
Department of Chemistry, GM University, Davangere, Karnataka, India

Department of Chemistry, GMIT, Visvesvaraya Technological University, Davangere, Karnataka, India

G. M. Bhavana · M. S. Kiran · P. Yuvarajagouda · N. Hareesha (✉)
Department of Chemistry, GM University, Davangere, Karnataka, India

B. P. Sanjay
Department of Chemical Engineering and Biotechnology, National Taipei University of Technology (Taipei Tech), Taipei, Taiwan

Department of Materials and Mineral Resources Engineering, National Taipei University of Technology, Taipei, Taiwan

R. Naik et al. (eds.), *Biocompatible Nanomaterials for Smart Sensors*, Nanostructure Science and Technology,
https://doi.org/10.1007/978-3-032-01745-1_5

Keywords Green synthesis · Nanocomposites · Heavy metal detection · Sustainable nanotechnology · Electrochemical sensors

5.1 Introduction

One ongoing and serious worldwide issue is the contamination of environmental matrices by heavy metals, especially in air, soil, and water [1]. Because of their acute toxicity, lack of biodegradability, and significant bio-accumulative potential, mercury (Hg), lead (Pb), and cadmium (Cd) are the most dangerous of the numerous hazardous metals [2, 3]. Neurotoxicity, nephrotoxicity, carcinogenicity, and developmental abnormalities are just a few of the serious health effects that can arise from long-term exposure to even minute amounts of these metals. For efficient monitoring and control, these hazards need the development of sensitive, dependable, and real-time detection methods [4].

The gold standards for detecting heavy metals have traditional analytical techniques such as X-ray fluorescence (XRF), inductively coupled plasma mass spectrometry (ICP-MS), and atomic absorption spectroscopy (AAS), but great operational costs, significant sample preparation, restricted field-deploy ability, and reliance on centralized laboratory infrastructure are some of the drawbacks of these methods, despite their great sensitivity and accuracy [5]. These difficulties highlight the pressing need for substitute detection techniques that are not only sensitive and effective but also portable and reasonably priced. However, conventional methods for synthesizing nanocomposites often involve toxic reagents, energy-intensive processes, and the generation of hazardous by-products, which are incompatible with the principles of green and sustainable chemistry. In recent years, nanomaterial-based sensors have emerged as promising platforms for environmental monitoring, offering enhanced sensitivity, selectivity, and the potential for miniaturization [6]. Among these, nanocomposites hybrid materials composed of at least one nanoscale component have demonstrated exceptional performance, thanks to their unique physicochemical properties.

Green synthesis techniques have attracted a lot of interest as environmentally friendly substitutes to allay these worries. By using natural reducing and stabilizing agents made from microorganisms, plant extracts, and biopolymers, these techniques do away with the need for dangerous chemicals. Moreover, the bioactive functional groups introduced during green synthesis can enhance metal ion binding and improve sensor performance. The integration of green chemistry principles into nanomaterial design not only supports sustainability goals but also aligns with the emerging paradigm of circular economy and eco-conscious innovation [7].

This chapter provides a comprehensive overview of green synthesized nanocomposites for the detection of mercury, lead, and cadmium. It emphasizes the intersection of green nanotechnology and smart sensing by discussing material design strategies, synthesis techniques, structural and functional characterization, and sensor performance. Furthermore, it presents illustrative case studies, evaluates current technological limitations, and outlines future directions for advancing eco-friendly sensor platforms in environmental monitoring.

5.2 Fundamentals of Green Nanocomposite Synthesis

The invent of nanotechnology has significantly transformed sensor expansion, especially in the field of environmental monitoring. However, synthesis methods for conventional nanocomposites such as chemical vapor deposition, sol-gel processes, and hydrothermal techniques are often connected with high energy demands, the use of toxic chemicals, and the generation of hazardous by-products [8]. These drawbacks highlight the crucial need for alternative synthesis strategies that are safer, more sustainable, and environmentally harmless. Green synthesis offers such a proposal, aligning with the principles of green chemistry and contributing to responsible transformation in nanomaterial science.

5.3 Principles of Green Chemistry in Nanomaterial Synthesis

Green nanotechnology has 12 Principles of Green Chemistry, particularly highlighting the prevention of waste, the use of nontoxic and renewable reagents, and energy efficiency, as illustrated in Fig. 5.1 [9]. These principles serve as a foundational framework for designing environmentally responsible nanocomposite synthesis methods.

- Environmentally harmless solvents (e.g., water, ethanol).
- Nontoxic reducing agents (e.g., plant polyphenols, amino acids).
- Energy-efficient processes, low-temperature.
- Renewable biogenic materials as reactants, capping, or stabilizing agents.

By integrating these practices, researchers can significantly reduce the ecological impact of nanomaterial production while improving the biocompatibility, selectivity, and functional efficiency of the resulting sensor platforms.

5.4 Natural Sources for Nanocomposite Synthesis

Biogenic resources including plants, microbes, algae, and biopolymers have emerged as effective and sustainable sources for nanocomposite synthesis [10].

- Plant extracts, rich in phytochemicals such as flavonoids, alkaloids, phenolics, and terpenoids, serve dual roles as reducing and stabilizing agents [11]. For example, *Azadirachta indica* (neem) and *Camellia sinensis* (green tea) extracts

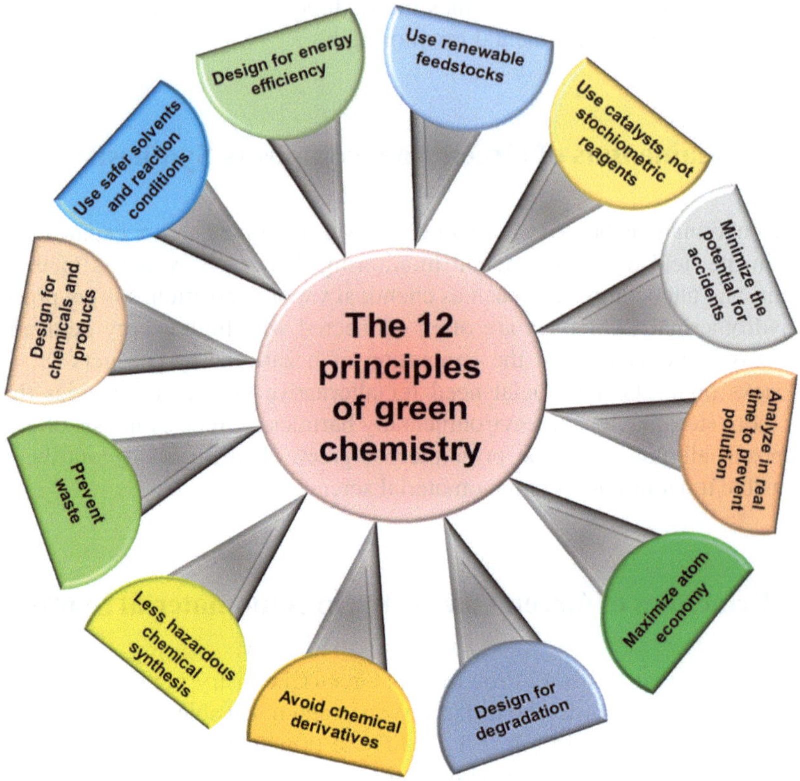

Fig. 5.1 The 12 principles of green chemistry adapted for nanomaterial synthesis [9]

have been successfully used in the synthesis of metal oxide nanoparticles [12, 13].

- Microbial agents, including bacteria (*Bacillus subtilis*, *Pseudomonas* spp.) and fungi (*Aspergillus flavus*, *Fusarium oxysporum*), can biosynthesize nanoparticles via enzymatic reduction pathways [14, 15].
- Biopolymers such as chitosan, cellulose, alginate, and starch provide both structural frameworks and functional groups (–OH, $-NH_2$, –COOH) that enhance metal ion binding and sensor efficiency [16]. These naturally derived sources often confer surface functionality, improving the selectivity and reactivity of the synthesized nanocomposites for heavy metal detection.

5.5 Green Synthesis Strategies

Green synthesis methodologies aim to eliminate hazardous reagents and reduce environmental impact by utilizing biological systems and low-energy processes [17]. These strategies rely on the natural reducing, stabilizing, and capping capabilities of plant extracts, microbes, and biopolymers to construct functional nanostructures under mild, aqueous conditions [18]. Table 5.1 summarizes common green synthesis routes for nanocomposite fabrication, highlighting their key characteristics and advantages for sustainable sensing applications.

These strategies provide a versatile platform for engineering nanocomposites with customized properties, ensuring both functional performance and environmental responsibility in sensor applications.

5.6 Composition and Structure Control

Controlling the composition and structural features of nanocomposites is crucial for optimizing their sensing performance [19]. Green synthesis offers a bio-assisted, self-regulating environment wherein natural extracts and biopolymers influence particle nucleation, growth, morphology, and surface functionalization [20, 21]. By carefully selecting precursors such as metal salts and pairing them with biogenic reducing/stabilizing agents (e.g., plant extracts or microbial enzymes), researchers can tailor the nanocomposite's elemental composition, crystallinity, and doping levels (e.g., N, S) [22]. Biopolymeric matrices like chitosan and cellulose further enhance stability and provide active sites for selective ion binding [16].

Key synthesis parameters such as pH, temperature, precursor-to-extract ratio, and reaction duration dictate the resulting nanoparticle size, shape, surface roughness, and porosity all of which impact the sensor's sensitivity and selectivity [23]. For instance, higher synthesis temperatures may yield larger, more crystalline particles, whereas acidic conditions may promote anisotropic growth, producing rod-like or flower-shaped nanostructures with increased surface area [24]. Advanced characterization techniques including X-ray diffraction (XRD), scanning electron microscopy (SEM), transmission electron microscopy (TEM), and Fourier-transform infrared spectroscopy (FTIR) are employed to investigate structural uniformity, crystallographic phases, functional groups, and elemental distributions [25–28]. These insights enable precise tuning of material properties to meet specific

Table 5.1 Summary of nanocomposite fabrication by green synthesis methods

Method	Key features	Advantages
Phytochemical reduction	Utilizes plant extracts to reduce metal salts	Eco-friendly, nontoxic, performed under mild conditions
Sol-gel method	Hydrolysis and condensation of metal precursors in aqueous media	Scalable, low-temperature, uniform material production
Hydrothermal synthesis	High-pressure reactions in aqueous environments	Crystallinity control, minimal chemical additives
Microwave-assisted synthesis	Rapid synthesis via green solvents under microwave irradiation	Energy-efficient, uniform particle nucleation
Bio-templating	Employs biological templates (e.g., leaves, microbial cell walls)	Biomimetic structure, enhanced surface reactivity

sensor requirements. Ultimately, the ability to control nanocomposite structure and composition through green synthesis opens new avenues for developing highly responsive and application-specific materials, particularly for detecting hazardous metals such as Hg^{2+}, Pb^{2+}, and Cd^{2+} in complex environmental systems [29, 30].

5.7 Properties of Green-Synthesized Nanocomposites

Green-synthesized nanocomposites have emerged as highly promising materials for heavy metal sensing due to their unique structural, morphological, and functional characteristics. These features many of which are intrinsically linked to the green synthesis route offer not only environmental compatibility but also performance advantages that make such materials well-suited for smart sensor applications targeting toxic metals like mercury (Hg^{2+}), lead (Pb^{2+}), and cadmium (Cd^{2+}) [31].

5.8 Structural and Morphological Features

The structural and morphological properties of nanocomposites are critical determinants of their sensing performance. Attributes such as particle size, shape, dispersion, porosity, and surface topology significantly influence surface area, mass transport dynamics, and interaction efficiency with target analytes [28]. One of the hallmark advantages of green synthesis is its biologically mediated control over these attributes, yielding nanocomposites with superior architectures for environmental monitoring applications [32].

Figure 5.2 shows the schematic representation of the structural and surface features of green-synthesized nanocomposites. This figure illustrates the process of green synthesis and the resulting nanocomposite features that enhance sensing performance. Typically, green-synthesized nanocomposites demonstrate narrow particle size distributions, uniform nanoparticle dispersion, and high surface-to-volume ratios [33, 34]. These features are largely attributed to the templating and stabilizing effects of plant-derived phytochemicals, microbial metabolites, or biopolymeric scaffolds rich in hydroxyl (–OH), carboxyl (–COOH), and amine ($–NH_2$) functional groups [34, 35]. These bioactive moieties serve dual roles by both preventing agglomeration and promoting the formation of mesoporous or microporous networks, which are ideal for enhancing the diffusion and adsorption of heavy metal ions [36].

Moreover, green synthesis routes often facilitate the formation of anisotropic nanostructures such as nanorods, nanoplates, nanosheets, and core–shell architectures [37]. This shape control is made possible by selective binding and growth inhibition along specific crystallographic directions, driven by natural capping agents (e.g., flavonoids, polysaccharides) [18]. These anisotropic structures are not

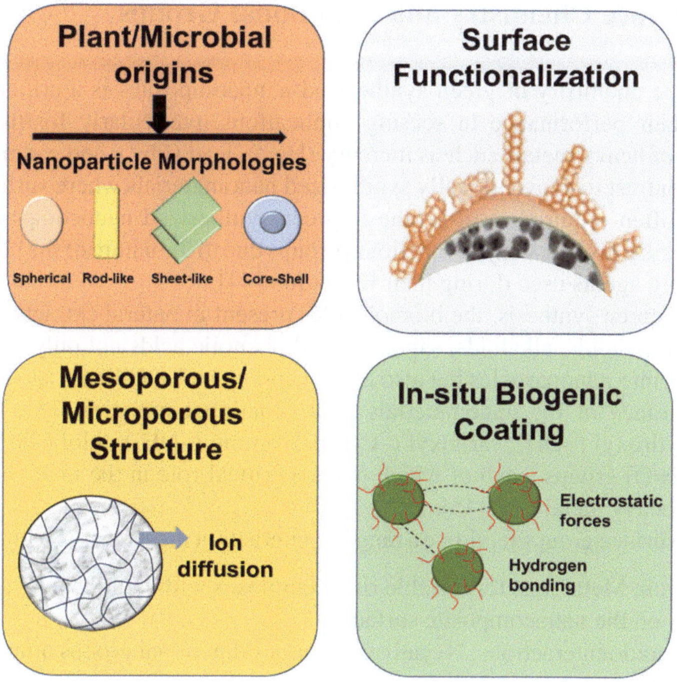

Fig. 5.2 Schematic representation of structural and surface features in green-synthesized nanocomposites. Diverse morphologies (e.g., spherical, rod, sheet, core–shell) arise through biogenic shape control. Surface functionalization where –OH, $-NH_2$, and –COOH groups prevents agglomeration and provides binding sites for heavy metal ions (Hg^{2+}, Pb^{2+}, Cd^{2+}). Porous networks support ion diffusion, while in situ organic coatings enable selective interaction via electrostatic forces, hydrogen bonding, and chelation

only morphologically distinctive but also exhibit superior catalytic and electrochemical activity due to the increased exposure of reactive facets.

Another key advantage of green synthesis is the in situ deposition of biogenic surface coatings during nanoparticle formation [38]. These coatings, typically rich in functional groups, confer multiple benefits: enhanced aqueous dispersibility, improved colloidal stability, and abundant binding sites for heavy metals via electrostatic interactions, chelation, and hydrogen bonding [39, 40]. These multifunctional surfaces significantly improve sensor responsiveness and specificity, enabling the detection of toxic metal ions even at trace concentrations. Importantly, the structural sophistication achieved through green synthesis is not merely incidental, it is actively engineered via bio-mediated control over nucleation, growth kinetics, and surface chemistry [41, 42]. As a result, green-synthesized nanocomposites often demonstrate superior structural uniformity, surface reactivity, and functional tunability, sometimes exceeding the performance of their conventionally synthesized counterparts.

5.9 Surface Chemistry and Functional Groups

The surface chemistry of green-synthesized nanocomposites is a critical determinant of their performance in sensing applications, particularly for the selective detection of heavy metals such as mercury (Hg^{2+}), lead (Pb^{2+}), and cadmium (Cd^{2+}) [43]. In contrast to conventionally synthesized nanomaterials where surface modification is often required, post-synthesis green-synthesized nanocomposites inherently possess a diverse array of functional groups due to the nature of the bio-reducing and capping agents used during their formation [44].

During green synthesis, the biomolecules present in natural extracts such as flavonoids, terpenoids, alkaloids, saponins, and phenolic acids not only reduce metal precursors into nanoparticles but also act as capping and stabilizing agents, decorating the surface of the nanomaterials with functional moieties [18]. These may include hydroxyl (–OH), carboxyl (–COOH), amino (–NH_2), thiol (–SH), and carbonyl (–C=O) groups, each of which plays a critical role in the material's interaction with heavy metal ions [45].

These surface groups facilitate a range of interaction mechanisms with metal ions:

- Chelation: Metal ions form stable ring complexes with multidentate or bidentate ligands on the nanocomposite surface.
- Electrostatic interactions: Negatively charged functional groups attract and bind positively charged metal ions.
- Hydrogen bonding: Further stabilizes the ion-nanocomposite interactions, contributing to strong and specific binding.
- Redox activity: In some systems, functional groups may participate in electron transfer reactions, enhancing electrochemical detection capabilities.

Such rich surface chemistry not only improves selectivity and affinity for target ions but also enhances key sensor parameters, including signal strength, detection limit, and response time. For example: Amine-rich surfaces show strong complexation with Pb^{2+} and Cd^{2+}. Thiol (–SH) groups selectively coordinate with Hg^{2+}, a soft metal ion, through highly polarizable sulfur-metal bonds [46].

Besides their broad functionality, these nature-derived surface groups impart notable environmental and operational improvements. By improving aqueous dispersibility, they better connect the sensor interface and analytes in aqueous media. The biocompatible and absence of toxicity further render green nanocomposites well-equipped for in situ sensing applications. Ultimately, the built-in functionalization provided by green synthesis forms the foundation of the material's sensing behavior. It ensures a synergistic blend of chemical reactivity, and mechanical stability, removing the need for additional surface functionalization or toxic chemicals. This bio-inspired surface architecture exemplifies green chemistry principles while enabling the development of highly selective and sustainable sensor platforms for practical environmental monitoring.

5.10 Electronic and Optical Properties

The electronic and optical properties of green-synthesized nanocomposites are pivotal to their performance as smart sensing platforms, particularly for the detection of heavy metals via electrochemical, colorimetric, and fluorescence-based techniques [47]. These functional attributes are inherently influenced by the composition, crystallinity, morphology, and surface chemistry of the nanomaterials factors that are, in turn, governed by the green synthesis route and the nature of the biological reducing and stabilizing agents employed.

5.10.1 Electronic Properties

Green-synthesized nanocomposites often exhibit enhanced charge transport characteristics, especially when incorporating conductive materials such as reduced graphene oxide (rGO), carbon nanotubes (CNTs), or metallic nanoparticles [34]. The presence of bio-derived capping agents and natural polymers can facilitate electron donor–acceptor interactions, thereby promoting efficient redox activity at the sensor interface. This is particularly advantageous in electrochemical sensors, where signal transduction depends on rapid and efficient electron transfer between the electrode and the analyte [48]. For instance, rGO synthesized using green tea extract offers not only a high surface area and conductivity but also an abundance of oxygen-containing functional groups [47]. These groups enhance the adsorption and redox interaction with metal ions such as Pb^{2+}, thereby improving the sensitivity of electrochemical detection. Likewise, gold and silver nanoparticles, synthesized using plant extracts, exhibit excellent electrical conductivity and electrocatalytic activity, making them effective amplifiers of electrochemical signals in voltametric or aerometric measurements [49].

5.10.2 Optical Properties

Green-synthesized nanocomposites also demonstrate unique and tuneable optical properties that are highly advantageous for colorimetric and fluorescence-based sensing. Metallic nanoparticles, such as Au and Ag, exhibit surface plasmon resonance (SPR)—a phenomenon where collective oscillation of electrons at the nanoparticle surface generates strong absorption peaks in the visible range. Upon interaction with heavy metal ions, these SPR peaks often shift in wavelength or intensity, providing a visual cue for detection. This colorimetric response can be observed by the naked eye or quantified using UV–Vis spectroscopy, enabling simple, cost-effective detection methods. In parallel, carbon-based fluorescent nanomaterials, such as carbon dots, synthesized from green sources like fruit peels or medicinal plants, display strong photoluminescence (PL) in the blue to green spectral range under UV light [50]. These fluorescent signals are highly sensitive to quenching by heavy metal ions such as $Hg2^{+}$ and Cd^{2+}, due to mechanisms involving photo-induced electron transfer (PET) or Förster resonance energy transfer (FRET) [51]. Changes in emission intensity or wavelength can thus serve as highly sensitive indicators for trace-level metal ion detection.

5.10.3 Tunability of Electronic and Optical Behavior

A unique advantage of green-synthesized nanomaterials is the tunability of their electronic and optical behavior through controlled synthesis parameters. Adjusting factors such as precursor concentration, reaction temperature, pH, and biogenic agent type allows for fine-tuning of critical properties including bandgap energy, conductivity, emission wavelength, plasmonic response. This adjustability enables the strategic design of nanocomposites for application-specific sensing, enabling scientists to tailor material characteristics to match the demands of various environmental or biological detection systems.

5.11 Biocompatibility and Environmental Safety

A major merit of green-synthesized nanocomposites is their natural biocompatibility and green approach, setting them apart from traditionally manufactured nanomaterials, where traditional methods often rely on toxic chemicals and generate toxic waste. This eco-friendly nature of green-synthesized nanocomposites is crucial when developing sensors for use in sensitive settings like freshwater habitats, biomedical fields, and agricultural zones, where reducing the risks to both human health and ecological systems.

Development of biocompatible nanocomposites from the use of natural, non-toxic reducing and stabilizing agents. Commonly used biological sources are micro-organisms, plant-based extracts, and biopolymers such as chitosan, alginate, and cellulose. These agents not only facilitate nanoparticle generation but also naturally functionalize the surfaces with bio-derived functional groups which reduce cytotoxicity and improve compatibility with living cells and biological tissues. For instance, silver nanoparticles produced with the help of chitosan have shown excellent biocompatibility in aquatic toxicity tests, supporting their use in real-time water quality monitoring without affecting aquatic organisms.

From an environmental perspective, green-synthesized nanocomposites provide advantages in biodegradability and minimal environmental persistence. In contrast to conventional nanomaterials stabilized with synthetic surfactants or nondegradable polymers, green nanocomposites, especially those obtained from renewable sources such as starch, fruit peels, or lignin, can decompose into safe organic residues under natural conditions [52]. This degradability reduces ecological stress and supports sustainable waste management, reducing the need for complex post-use recycling or toxic disposal procedures. Additionally, the control of toxic chemical reagents such as $NaBH_4$, N_2H_4, and volatile organic solvents during the synthesis process greatly reduces the overall toxicity profile of the final nanomaterial [53]. This not only makes the materials easier to manage during production and deployment but also aligns with the principles of green chemistry, which emphasize less risky, sustainability, and minimal environmental impact. In summary, the green synthesis approach offers a dual benefit—it ensures high-performance sensing functionality while maintaining a strong commitment to human health and environmental protection. These attributes make green-synthesized nanocomposites particularly attractive for real-world sensing applications, especially where sustainability and safety are paramount.

5.12 Comparative Summary

Green-synthesized nanocomposites deliver a range of constructional, chemical, and functional benefits that make them highly adaptable for sustainable sensor applications. Unlike traditional nanomaterials, which are often synthesized using high-consumption processes and hazardous chemicals, green nanocomposites are typically produced under non-extreme conditions using organic reducing and stabilizing agents. Consequently, they show exceptional ecological alignment, intrinsic surface functionalization, and improve biocompatibility, making them especially valuable for applications in environmental and biological matrices.

In-addition, green synthesis techniques empower simplified and cost-effective fabrication, often using renewable or waste organic matter, while maintaining or even improving upon the performance characteristics critical for heavy metal detection such as susceptibility, precision, and durability. Their ability to operate effectively in field-deployable formats also makes them ideal candidates for real-time, on-site monitoring systems. Table 5.2 Illustrates a detailed comparison between green-synthesized and conventional nanocomposites, highlighting their differences in synthesis strategy, toxicity, scalability, and suitability for practical sensing applications.

5.13 Mechanisms of Heavy Metal Detection

5.13.1 Electrochemical Sensing

Electrochemical sensing remains one of the most powerful and widely employed strategies for the detection of toxic heavy metals, owing to its exceptional sensitivity, selectivity, and compatibility with portable, low-cost instrumentation. In the

Table 5.2 Comparative evaluation of green-synthesized and conventional nanocomposites

Property	Green-synthesized nanocomposites	Conventional nanocomposites
Synthesis method	Low-temperature, eco-friendly, bio-based	High-energy, chemical-intensive
Reducing/capping agents	Plant extracts, microbes, biopolymers	Toxic chemicals (e.g., $NaBH_4$, hydrazine
Surface functionalization	Inherent from biogenic capping agents (e.g., –OH, –COOH, $-NH_2$)	Often requires post-synthesis modification
Toxicity	Minimal; safe for aquatic, soil, and biological applications	Potentially toxic by-products and residues
Morphological control	Naturally guided crystal growth; anisotropic structures possible	Requires precise control of surfactants and conditions
Biocompatibility	High; suitable for biosensors and environmental sensors	Moderate to low; concerns in biological environmental exposure
Degradability	Biodegradable components; low environmental impact	Non-biodegradable; disposal and persistence issues
Cost and scalability	Low-cost, simple synthesis; scalable using biomass or waste	Expensive reagents and infrastructure needed
Field application readiness	Portable, stable, and adaptable for resource-limited or remote settings	Typically, lab-bound; not always optimized for real-time or in-situ use

context of green nanotechnology, green-synthesized nanocomposites have been instrumental in advancing this field, providing not only environmentally friendly materials but also superior sensing performance through enhanced surface functionality and charge transport properties.

The core principle of electrochemical detection involves the interaction between target metal ions and a nanocomposite-modified electrode surface, which leads to measurable changes in electrochemical signals. When green-synthesized nanocomposites such as metallic nanoparticles, reduced graphene oxide (rGO), or carbon nanotubes embedded in bio-derived matrices like chitosan, cellulose, or alginate are deposited onto electrodes (e.g., glassy carbon or screen-printed electrodes), they create a highly reactive interface capable of mediating efficient electron transfer processes.

The functional groups introduced during green synthesis such as hydroxyl (–OH), carboxyl (–COOH), and amino ($-NH_2$) moieties play a crucial role by facilitating metal ion complexation, electrostatic attraction, and surface adsorption. These interactions not only improve selectivity toward metal ions like Pb^{2+}, Cd^{2+}, and Hg^{2+}, but also enhance the kinetics of redox reactions, which are central to electrochemical sensing. Upon exposure to a metal-contaminated solution, the target ions diffuse to the electrode surface and bind to active sites on the nanocomposite. This interaction alters the interfacial charge transfer dynamics, which can be probed using electroanalytical techniques, such as cyclic voltammetry (CV), square wave voltammetry (SWV), and differential pulse voltammetry (DPV).

In CV, for example, the formation of a metal–ligand complex may shift the peak potential or modify the peak current, providing a quantitative signal that correlates with metal ion concentration [54]. These shifts are governed by both the kinetic parameters (e.g., electron transfer rate) and thermodynamic properties (e.g., standard electrode potentials) of the analyte nanocomposite system [55]. Moreover, the use of high-conductivity nanostructures such as rGO, especially when synthesized via plant extract-mediated reduction, introduces defect sites and enhances electron mobility and adsorption capacity. A representative example includes a chitosan–rGO nanocomposite synthesized using *Camellia sinensis* (green tea) extract, wherein plant-derived flavonoids act as both reducing and stabilizing agents [56]. When incorporated into a sensor platform, this nanocomposite demonstrated ultrasensitive detection of Pb^{2+} ions, achieving a nanomolar detection limit. This performance was attributed to the synergistic interplay between chitosan's amine-rich structure and rGO's high conductivity, along with the green-derived surface functionalities that selectively bind lead ions. Such innovations illustrate how green-synthesized nanocomposites can be seamlessly integrated into electrochemical sensors to meet modern requirements for sustainability, portability, and high analytical performance. These sensors are particularly well-suited for in-field environmental monitoring, offering a compelling solution for rapid and eco-friendly detection of heavy metal contaminants.

5.13.2 Colorimetric Sensing

Colorimetric sensing is a highly intuitive and practical analytical approach that detects target analytes based on visible color changes, either observed by the naked eye or measured spectrophotometrically [57]. Owing to its simplicity, low cost, and instrument-free usability, colorimetric sensing is especially attractive for on-site and point-of-care applications making it ideal for the detection of toxic heavy metals in environmental and biological samples [58]. In the realm of nanotechnology, colorimetric sensors frequently leverage the localized surface plasmon resonance (LSPR) of metallic nanoparticles, particularly those of gold (AuNPs) and silver (AgNPs). These nanoparticles exhibit strong absorption in the visible region, which shifts upon changes in particle size, shape, or interparticle distance providing a direct optical readout of analyte interactions.

Green-synthesized nanocomposites, especially those derived from plant extracts or microbial systems, have gained significant attention for colorimetric applications due to their inherent environmental safety and functional surface chemistries. During green synthesis, phytochemicals such as flavonoids, polyphenols, and terpenoids serve as both reducing agents and stabilizers, producing nanoparticles coated with reactive functional groups like hydroxyl (–OH), carbonyl (C=O), and amine ($-NH_2$). These groups facilitate selective interactions with metal ions, enabling colorimetric detection with high specificity. The fundamental detection mechanism typically involves nanoparticle aggregation triggered by metal ion binding. Upon the introduction of target ions (e.g., Hg^{2+}, Pb^{2+}, Cd^{2+}), the nanoparticles may aggregate due to cross-linking via metal ion bridges. Undergo surface charge neutralization, destabilizing the colloid. Experience oxidation or morphological transformation, altering optical properties. These changes lead to a red shift and broadening of the LSPR band, producing a distinct color change. For example, dispersed AuNPs appear red (~520 to 530 nm LSPR). Upon aggregation, the solution turns blue or purple due to shifted LSPR peaks [59]. In some systems, metal ions induce direct oxidation of the nanoparticles. Mercury ions (Hg^{2+}), for instance, bind strongly to thiol or sulfur-containing groups, facilitating the oxidation of AgNPs. This interaction may lead to a loss of color or a shift from pale yellow to brown, as seen in green-synthesized silver nanoparticle systems.

A representative example includes AgNPs synthesized using *Ocimum sanctum* (holy basil) leaf extract. The phytochemicals in the extract reduce silver ions and cap the resulting nanoparticles [60]. Upon exposure to Hg^{2+}, the solution rapidly changes from pale yellow to brown—indicative of nanoparticle aggregation and redox interaction. This qualitative visual cue is further confirmed by UV–Vis spectroscopy, which records a shift in the LSPR peak correlating with metal ion concentration, allowing semi-quantitative or quantitative analysis. Colorimetric sensors based on green nanomaterials are particularly suited for resource-limited environments, where portability, disposability, and rapid results are critical. These sensors can be easily integrated into paper-based analytical devices (PADs) or lateral flow assays, further extending their field-deployable capabilities [61].

Moreover, the biocompatibility and nontoxic nature of green-synthesized nanocomposites minimize environmental impact, making them safe for applications in ecosystems, agriculture, and biomedical settings. While the visual simplicity of colorimetric sensing is a major advantage, potential challenges include interference from colored backgrounds or matrix effects in complex samples. However, these issues can be mitigated by tailoring the surface chemistry of nanoparticles by employing specific ligands or biorecognition elements (e.g., aptamers, peptides). Such enhancements improve selectivity, stability, and reliability, making green nanocomposite-based colorimetric sensors a compelling choice for sustainable and sensitive heavy metal detection [62].

5.13.3 Fluorescence-Based Detection

Fluorescence-based sensing offers an exceptionally sensitive, selective, and noninvasive approach for the detection of heavy metal ions, often capable of detecting trace-level contaminants in complex environmental and biological matrices [63]. This technique relies on the modulation of a material's fluorescent properties typically through quenching or enhancement upon interaction with a specific analyte. In this context, green-synthesized fluorescent nanocomposites combine high photostability with eco-friendliness and inherent surface functionality, making them particularly attractive for deployment in smart sensor platforms [64]. A wide range of fluorescent nanomaterials, such as carbon quantum dots (CQDs), graphene quantum dots (GQDs), and rare-earth-doped oxides, have been synthesized using green methods from natural sources such as fruit peels, herbal extracts, and biopolymers [65]. These carbon-based nanostructures, typically smaller than 10 nm, are rich in functional groups hydroxyl (–OH), amino ($–NH_2$), and carboxyl (–COOH), which contribute to their aqueous solubility, biocompatibility, and metal ion binding capacity. In addition to reducing the environmental footprint, green synthesis often results in nanomaterials with high quantum yields, narrow emission spectra, and excellent optical stability.

Fluorescence-based detection of heavy metals typically proceeds via one of two pathways. Fluorescence Quenching: This is the most common mechanism, wherein the binding of metal ions such as Cd^{2+} or Hg^{2+} to the fluorescent nanomaterial leads to a decrease in photoluminescence intensity [51]. Quenching may occur through photoinduced electron transfer (PET), fluorescence resonance energy transfer (FRET), or inner filter effect (IFE). For instance, the high electron affinity of Hg^{2+} allows it to efficiently accept electrons from the excited state of a carbon dot, dissipating the energy via nonradiative pathways and thus suppressing fluorescence.

Fluorescence Enhancement: In some systems, binding of metal ions induces a conformational change or increases the rigidity of the fluorescent framework, thereby promoting radiative decay and enhancing emission [66]. Though less commonly observed, this mechanism offers unique opportunities for signal amplification. Figure 5.3 shows the variety of natural sources and synthetic approaches used to create nanosensors, including electrochemical, colorimetric, and fluorescence-based methods.

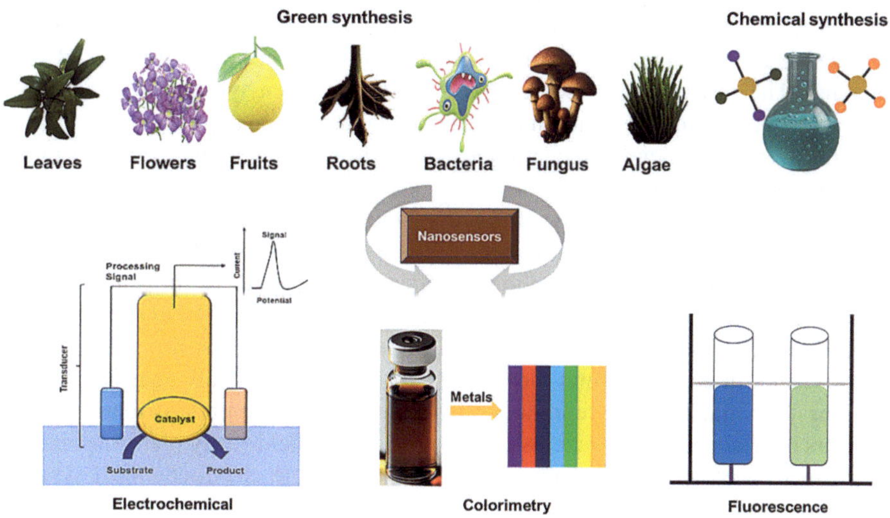

Fig. 5.3 Schematic representation of metal ions detection using green-synthesized nanocomposites

5.14 Case Studies and Applications

The practical application of green-synthesized nanocomposites in the detection of heavy metals has demonstrated significant promise across diverse environmental and biological systems. Numerous studies validate their sensitivity, selectivity, and eco-compatibility, highlighting their potential for real-world monitoring of toxic ions such as mercury (Hg^{2+}), lead (Pb^{2+}), and cadmium (Cd^{2+}) in matrices including water, soil, and bodily fluids.

The practical use of green-synthesized nanocomposites in heavy metal detection was comprehensively investigated in a work by Rabiaa Helim. These nanocomposites have demonstrated great potential in a variety of biological and environmental processes. The sensitivity, selectivity, and eco-compatibility of these nanocomposites have been confirmed by several studies, and they are essential for monitoring harmful ions like mercury (Hg^{2+}), lead (Pb^{2+}), and cadmium (Cd^{2+}) in matrices including water, soil, and body fluids in the real world. In keeping with the rising need for environmentally friendly solutions in environmental monitoring and public health protection, this study highlights the promise of green-synthesized nanocomposites as a sustainable and efficient heavy metal detection option [67].

The practical use of green-synthesized nanocomposites in heavy metal detection was comprehensively investigated in a work by Aayushi Kundu. Using a simple microwave technique without any chemicals, scientists created green-emissive carbon quantum dots (CQDs) from orange pomace, a biomass-based precursor. Excellent photostability, water solubility, and a high fluorescent quantum yield of 54.26% were all demonstrated by these CQDs. With detection limits of 59.6 nM and 14 nM, respectively, the CQDs demonstrated encouraging results for the detection of Cr^{6+} ions and 4-nitrophenol (4-NP). The study showed that the CQDs may be utilized to detect these analytes in the nanomolar range in a sensitive and selective manner. The research underscores the potential of green-synthesized CQDs as a sustainable and effective solution for heavy metal detection, aligning with the growing need for eco-friendly technologies in environmental monitoring and public health protection [67].

Table 5.3 depicts the utility of green-synthesized nanocomposites in detecting heavy metals. It highlights various sensors, including those based on AgNPs, CQDs, and carbon paste electrodes, and their applications across different water samples. The detection methods encompass colorimetric (including LSPR and UV-Vis), electrochemical, and fluorescence techniques, underlining the versatility of these green nanosensors in environmental monitoring. Although some challenges persist such as long-term stability and reproducibility under variable conditions many of these sensors have already been successfully integrated into field-ready platforms, including paper-based analytical devices (PADs) and portable electrochemical systems. Their deployment in post-mining areas, agricultural zones, and water treatment facilities underscores the increasing relevance of sustainable sensing technologies in decentralized environmental monitoring frameworks.

Table 5.3 Real-world applications of green-synthesized nanocomposite sensors for heavy metal detection

Target ion	Green nanocomposite sensor	Sample matrix	Detection method	Reference
Hg^{2+}	AgNPs synthesized using *Foeniculum vulgare* seed extract	Distilled water, drinking water, tap water, treated tap water	Colorimetric (LSPR shift)	[68]
Cd^{2+}	Carbon paste electrode based on biomass-derived orange peel activated carbon (ACOP) and multiwalled carbon nanotubes (MWCNTs)	River water samples (Rimac River, Peru)	Electrochemical	[69]
Cd^{2+}	Carbon duantum dots (CQDs) synthesized hydrothermally from *Prunus armeniaca*	Tap water, river water, laboratory water samples	Fluorescence	[70]
Hg^{2+}	Nitrogen-doped carbon quantum dots synthesized from citrus lemon juice	Tap water, packaging drinking water	Fluorescence	[71]
Hg^{2+}	L-tyrosine-stabilized AgNPs and AuNPs aqueous	Tap and drinking water	Colorimetric (UV-Vis)	[31]
Pb^{2+}	L-tyrosine-stabilized AuNPs	Aqueous, tap, and drinking water	Colorimetric (UV-Vis)	[31]

5.15 Future Perspectives and Challenges

The field of green-synthesized nanocomposites for heavy metal detection is undergoing rapid advancement and holds tremendous promise to redefine how environmental and public health monitoring is conducted. As the demand for real-time, portable, and sustainable sensing technologies continues to rise, green nanotechnology is well-positioned to drive the development of the next generation of smart, eco-friendly sensors.

One of the most promising directions is the integration of green nanocomposites with digital technologies, such as the Internet of Things (IoT), wireless communication modules, and smartphone-based analytical platforms. These innovations have the potential to enable continuous and remote monitoring of toxic heavy metals in systems such as municipal water supplies, agricultural runoff networks, and industrial wastewater streams, providing real-time alerts when contaminant levels exceed regulatory thresholds. Another emerging trend is the development of multi-analyte sensor systems capable of simultaneously detecting several heavy metals such as Hg^{2+}, Pb^{2+}, and Cd^{2+} without significant cross-reactivity. Achieving this functionality requires sophisticated nanostructure engineering, surface functionalization, and the use of hybrid nanomaterials that can selectively and independently interact with specific target ions.

Despite these advancements, several key challenges remain:

- Reproducibility and scalability of green synthesis methods continue to be major concerns. The inherent variability in natural precursors (e.g., plant extracts, microbial cultures) can lead to inconsistencies in nanoparticle size, morphology, and surface chemistry, which affects sensor reliability and performance [72].
- Standardization and regulatory approval are essential for the broader adoption of green nanocomposite-based sensors in environmental, industrial, and clinical settings [73]. Establishing validated protocols for sensor calibration, performance testing, and quality control is critical for commercial translation.
- Long-term operational stability under field conditions (e.g., high humidity, temperature fluctuations, UV exposure) remains a technical bottleneck. Many green-synthesized sensors still require optimization for storage, transport, and reuse in variable and sometimes harsh environmental conditions [33, 74].

Addressing these challenges will require multidisciplinary collaboration across domains including materials science, analytical chemistry, environmental engineering, biomedicine, and data science. Research efforts should focus on:

- Improving the consistency of green synthesis protocols using bioinformatics and metabolomics to standardize plant extract composition [75].
- Enhancing sensor durability via surface passivation or encapsulation without compromising eco-safety [76].
- Integrating artificial intelligence (AI) and machine learning (ML) for data processing and predictive modeling in sensor networks [77].

As the field continues to mature, the convergence of green nanotechnology, smart electronics, and sustainable innovation will play a vital role in enabling accessible, real-time, and environmentally responsible monitoring of toxic heavy metals supporting global goals in public health, resource management, and sustainable development.

5.16 Conclusion

Green-synthesized nanocomposites represent a groundbreaking and sustainable approach in the design of smart sensors for detecting highly toxic heavy metals such as mercury (Hg^{2+}), lead (Pb^{2+}), and cadmium (Cd^{2+}). By leveraging eco-friendly, renewable, and biologically derived resources, these materials offer a powerful synergy between high-performance sensing and environmental sustainability, a dual imperative in today's landscape of global pollution and public health challenges. This chapter has explored the fundamentals of green synthesis, the structural and functional properties of green nanocomposites, and their deployment across electrochemical, colorimetric, and fluorescence-based detection platforms. Case studies have demonstrated their effective use in diverse sample types, including water, soil, and biological fluids, demonstrating immense potential for field-based, cost-efficient, and highly specific heavy metal detection. Compared to conventional detection methods, green nanocomposite-based sensors not only achieve comparable sensitivity and specificity, but also provide distinct advantages in portability, biocompatibility, and environmental safety. These features make them ideal candidates for use in limited resources settings, real-time environmental surveillance, and noncentralized monitoring systems. Expecting, the future of green nanocomposite sensors lies in their combination with emerging technologies including Internet of Things (IoT) systems, multi-component sensing, and smartphone-based platforms to enable automated, continuous, and remote monitoring. Overcoming existing obstacles related to scalability, uniformity, sensor stability, and compliance standardization will be essential for facilitating their shift from research stage to widespread practical deployment. As global objectives continue to align with circular economy principles and sustainable innovation, green nanocomposites are poised to become indispensable tools for building a cleaner, healthier, and more durable world. These materials barely transcend scientific curiosity, symbolizing the emerging wave of smart environmental detection, bridging the gap between scientific advancement and ecological responsibility.

References

1. Buckner CA, Lafrenie RM et al (2016) IntechOpen, the world's leading publisher of open access books built by scientists, for scientists top 1%. INTECH 11:13
2. Edo GI, Samuel PO, Oloni GO, Ezekiel GO, Ikpekoro VO, Obasohan P, Ongulu J, Otunuya CF, Opiti AR, Ajakaye RS, Essaghah AEA, Agbo JJ (2024) Environmental persistence, bioaccumulation, and ecotoxicology of heavy metals. Chem Ecol 40(3):322–349. https://doi.org/10.1080/02757540.2024.2306839
3. Tolkou AK, Toubanaki DK, Kyzas GZ (2023) Detection of arsenic, chromium, cadmium, lead, and mercury in fish: effects on the sustainable and healthy development of aquatic life and human consumers. Sustain 15(23):16242. https://doi.org/10.3390/su152316242
4. Ramakrishnan J, Liu T, Zhang F, Seshadri K, Yu R (1955) Creative, U Pr Ep Rin t n Ot Pe Er Re v Iew Pe Er Re v Pr Ep Rin t n Ot Iew pp 1–34
5. Adesina KE, Burgos CJ, Grier TR, Sayam ASM, Specht AJ (2025) Ways to measure metals: from ICP-MS to XRF. Curr Environ Heal Rep 12(1):7. https://doi.org/10.1007/s40572-025-00473-y
6. Gudja NC, Munteanu FD (2024) Hybrid nanomaterials: a brief overview of versatile solutions for sensor technology in healthcare and environmental applications. Biosensors 14(2):67. https://doi.org/10.3390/bios14020067
7. Kucuk I, Sadak S, Silah H, Dindar CK, Bozal-Palabiyik B, Uslu B (2025) Enhanced electrocatalytic performance of eco-friendly nanomaterial-based sensors for the analysis of pharmaceutically important compounds. Top Catal. https://doi.org/10.1007/s11244-025-02074-3
8. Kharissova OV, Kharisov BI, González CMO, Méndez YP, López I (2019) Greener synthesis of chemical compounds and materials. R Soc Open Sci 6:191378. https://doi.org/10.1098/rsos.191378
9. Kurul F, Doruk B, Nur S (2025) Principles of green chemistry : building a sustainable future. Discov Chem. https://doi.org/10.1007/s44371-025-00152-9
10. Biogenic nanomaterials for environmental sustainability: principles .... https://books.google.co.in/books?hl=en&lr=&id=3DvsEAAAQBAJ&oi=fnd&pg=PR5&dq=Biogenic+resources+i

ncluding+plants,+microbes,+algae,+and+biopolymers+have+emerged+as+effective+and+sustainable+sources+for+nanocomposite+synthesis&ots=_ht_Un6ey&sig=xcNruyvOkdZrc89tz6CQfGQHbsk&redir_esc=y#v=onepage&q&f=false. Accessed 23 April 2025

11. Ovais M, Khalil AT, Islam NU, Ahmad I, Ayaz M, Saravanan M, Shinwari ZK, Mukherjee S (2018) Role of plant phytochemicals and microbial enzymes in biosynthesis of metallic nanoparticles. Appl Microbiol Biotechnol 102(16):6799–6814. https://doi.org/10.1007/s00253-018-9146-7
12. Patil SP, Chaudhari RY, Nemade MS (2022) Azadirachta indica leaves mediated green synthesis of metal oxide nanoparticles: a review. Talanta Open 5:100083. https://doi.org/10.1016/j.talo.2022.100083
13. Al-Ghamdi SA, Alkathiri TA, Alfarraj AE, Alatawi OM, Alkathiri AS, Panneerselvam C, Vanaraj S, Darwish AAA, Hamdalla TA, Pasha A, Khasim S (2022) Green synthesis and characterization of zinc oxide nanoparticles using camellia sinensis tea leaf extract and their antioxidant, anti-bactericidal and anticancer efficacy. Res Chem Intermed 48(11):4769–4783. https://doi.org/10.1007/s11164-022-04845-z
14. Qamar SUR, Ahmad JN (2021) Nanoparticles: mechanism of biosynthesis using plant extracts, bacteria, fungi, and their applications. J Mol Liq 334:116040. https://doi.org/10.1016/j.molliq.2021.116040
15. Rai M, Bonde S, Golinska P, Trzcińska-Wencel J, Gade A, Abd-Elsalam K, Shende S, Gaikwad S, Ingle AP (2021) Extracellular biosynthesis of selenium nanoparticles using some species of lactobacillus fusarium as a novel fungus for the synthesis of nanoparticles: mechanism and applications. J Fungi 7(2):1–24
16. Khademian E, Salehi E, Sanaeepur H, Galiano F, Figoli A (2020) A systematic review on carbohydrate biopolymers for adsorptive remediation of copper ions from aqueous environments-part a: classification and modification strategies. Sci Total Environ 738:139829. https://doi.org/10.1016/j.scitotenv.2020.139829
17. Singh J, Dutta T, Kim KH, Rawat M, Samddar P, Kumar P (2018) "Green" synthesis of metals and their oxide nanoparticles: applications for environmental remediation. J Nanobiotechnol 16(1):1–24. https://doi.org/10.1186/s12951-018-0408-4
18. El-Seedi HR, El-Shabasy RM, Khalifa SAM, Saeed A, Shah A, Shah R, Iftikhar FJ, Abdel-Daim MM, Omri A, Hajrahand NH, Sabir JSM, Zou X, Halabi MF, Sarhan W, Guo W (2019) Metal nanoparticles fabricated by green chemistry using natural extracts: biosynthesis, mechanisms, and applications. RSC Adv 9(42):24539–24559. https://doi.org/10.1039/c9ra02225b
19. Said RAM, Hasan MA, Abdelzaher AM, Abdel-Raoof AM (2020) Review—insights into the developments of nanocomposites for its processing and application as sensing materials. J Electrochem Soc 167(3):037549. https://doi.org/10.1149/1945-7111/ab697b
20. Jones OG, McClements DJ (2010) Functional biopolymer particles: design, fabrication, and applications. Compr Rev Food Sci Food Saf 9(4):374–397. https://doi.org/10.1111/j.1541-4337.2010.00118.x
21. Menichetti A, Mavridi-Printezi A, Mordini D, Montalti M (2023) Effect of size, shape and surface functionalization on the antibacterial activity of silver nanoparticles. J Funct Biomater 14(5):244. https://doi.org/10.3390/jfb14050244
22. Ullah Z, Gul F, Iqbal J, Abbasi BA, Kanwal S, Chalgham W, El-Sheikh MA, Diltemiz SE, Mahmood T (2023) Biogenic synthesis of multifunctional silver oxide nanoparticles (Ag2ONPs) using parieteria alsinaefolia delile aqueous extract and assessment of their diverse biological applications. Microorganisms 11(4):69. https://doi.org/10.3390/microorganisms11041069
23. Khan MI, Mukherjee K, Shoukat R, Dong H (2017) A review on PH sensitive materials for sensors and detection methods. Microsyst Technol 23(10):4391–4404. https://doi.org/10.1007/s00542-017-3495-5
24. Li Y, Shen W (2014) Morphology-dependent nanocatalysts: rod-shaped oxides, vol 43. RSC Publishing, London. https://doi.org/10.1039/c3cs60296f
25. Prasad RD, Prasad RS, Prasad RB, Prasad SR, Singha SB, Singha AD, Prasad RJ, Teli SB, Sinha P, Vaidya AK, Saxena S, Saxena UR, Harale A, Deshmukh MB, Padvi MN, Navathe

GJ (2024) A review on modern characterization techniques for analysis of nanomaterials and biomaterials. ES Energy Environ 23:1087. https://doi.org/10.30919/esee1087
26. Varun DN, Manjunatha JG, Hareesha N, Sandeep S, Mallu P, Karthik CS, Prinith NS, Sreeharsha N, Asdaq SMB (2021) Simple and sensitive electrochemical analysis of riboflavin at functionalized carbon nanofiber modified carbon nanotube sensor. Monatshefte Chem 152(10):1183–1191. https://doi.org/10.1007/s00706-021-02839-y
27. Donnakatte Neelalochana V, Tomasino E, Di Maggio R, Cotini O, Scardi P, Mammi S, Ataollahi N (2023) Anion exchange membranes based on chemical modification of recycled PET bottles. ACS Appl Polym Mater 5(9):7548–7561. https://doi.org/10.1021/acsapm.3c01391
28. Donnakatte Neelalochana V, Mancini I, Loi N, Cufalo G, D'Anzi A, Scardi P, Ataollahi N (2025) Sustainable conversion of PET waste bottle into proton exchange membranes for fuel cells. ACS Appl Energy Mater. https://doi.org/10.1021/acsaem.4c03313
29. Gnanaprakasam P, Jeena SE, Premnath D, Selvaraju T (2016) Simple and robust green synthesis of au NPs on reduced graphene oxide for the simultaneous detection of toxic heavy metal ions and bioremediation using bacterium as the scavenger. Electroanalysis 28(8):1885–1893. https://doi.org/10.1002/elan.201600002
30. Hareesha N, Soumya DM, Mounesh, Manjunatha JG, Rohit RN, Manikanta P, Varun DN, Ataollahi N, Thippeswamy BA, Pramoda K, Nagaraja BM (2024) Honeycomb polypore biomass-derived activated porous carbon nanosheets/graphite/nafion composite: green and sensitive electrocatalyst for nanomolar detection of Hg2+ ions and water-splitting reactions. J Environ Chem Eng 12(5):113584. https://doi.org/10.1016/j.jece.2024.113584
31. Annadhasan M, Muthukumarasamyvel T, Sankar Babu VR, Rajendiran N (2014) Green synthesized silver and gold nanoparticles for colorimetric detection of Hg2+, Pb2+, and Mn2+ in aqueous medium. ACS Sustain Chem Eng 2(4):887–896. https://doi.org/10.1021/sc400500z
32. Manik UP, Nande A, Raut S, Dhoble SJ (2020) Green synthesis of silver nanoparticles using plant leaf extraction of artocarpus heterophylus and azadirachta indica. Results Mater 6(March):100086. https://doi.org/10.1016/j.rinma.2020.100086
33. Pechyen C, Tangnorawich B, Toommee S, Marks R, Parcharoen Y (2024) Green synthesis of metal nanoparticles, characterization, and biosensing applications. Sensors Int 5(March):100287. https://doi.org/10.1016/j.sintl.2024.100287
34. Safajou H, Mizwari ZM, Rostaminia A, Khojasteh H, Aspoukeh P, Mazhari MP (2024) Green synthesis and enhanced photocatalytic performance of RGO/ZnO/Fe_3O_4 nanocomposites: a sustainable approach to environmental remediation. J Fluoresc. https://doi.org/10.1007/s10895-024-04014-y
35. Scotti L, Jaime Bezerra Mendonca F, Rodrigo Magalhaes Moreira D, Sobral da Silva M, Pitta I, Tullius Scotti M (2013) SAR, QSAR and docking of anticancer flavonoids and variants: a review. Curr Top Med Chem 12(24):2785–2809. https://doi.org/10.2174/1568026611212240007
36. Hassan M, Liu Y, Naidu R, Du J, Qi F, Donne SW, Islam MM (2021) Mesoporous biopolymer architecture enhanced the adsorption and selectivity of aqueous heavy-metal ions. ACS Omega 6(23):15316–15331. https://doi.org/10.1021/acsomega.1c01642
37. Madani M, Hosny S, Alshangiti DM, Nady N, Alkhursani SA, Alkhaldi H, Al-Gahtany SA, Ghobashy MM, Gaber GA (2022) Green synthesis of nanoparticles for varied applications: green renewable resources and energy-efficient synthetic routes. Nanotechnol Rev 11(1):731–759. https://doi.org/10.1515/ntrev-2022-0034
38. Popescu M, Ungureanu C (2023) Biosensors in food and healthcare industries: bio-coatings based on biogenic nanoparticles and biopolymers. Coatings 13(3):486. https://doi.org/10.3390/coatings13030486
39. Singh P, Garg A, Pandit S, Mokkapati VRSS, Mijakovic I (2018) Antimicrobial effects of biogenic nanoparticles. Nano 8(12):1–19. https://doi.org/10.3390/nano8121009
40. Schubert J, Chanana M (2018) Coating matters: review on colloidal stability of nanoparticles with biocompatible coatings in biological media, living cells and organisms. Curr Med Chem 25(35):4553–4586. https://doi.org/10.2174/0929867325666180601101859

41. Kamaraj SK, Thirumurugan A, Dhanabalan SS, Verma SK, Shajahan S (2024) Sustainable green synthesised nano-dimensional materials for energy and environmental applications. Routledge, London. https://doi.org/10.1201/9781003362241
42. Sagadevan S, Imteyaz S, Murugan B, Anita Lett J, Sridewi N, Weldegebrieal GK, Fatimah I, Oh WC (2022) A comprehensive review on green synthesis of titanium dioxide nanoparticles and their diverse biomedical applications. Green Process Synth 11(1):44–63. https://doi.org/10.1515/gps-2022-0005
43. Kiran V, Harini K, Thirumalai A, Girigoswami K, Girigoswami A (2024) Nanostructured carbon dots as ratiometric fluorescenrulers for heavy metal detection. Int J Nano Dimens 15(4):1–31. https://doi.org/10.57647/j.ijnd.2024.1504.26
44. Chaudhary M, Singh A (2024) Synthesis of nanoparticles using fruit waste and its pharmacological & catalytic applications: a review. BioNanoScience 14:3830–3845. https://doi.org/10.1007/s12668-024-01550-6
45. Bakhtiari S, Salari M, Shahrashoub M, Zeidabadinejad A, Sharma G, Sillanpää M (2024) A comprehensive review on green and eco-friendly nano-adsorbents for the removal of heavy metal ions: synthesis, adsorption mechanisms, and applications, vol 10. Springer, Cham. https://doi.org/10.1007/s40726-023-00290-7
46. Ding X, Hua Y, Chen Y, Zhang C, Kong X (2015) Heavy metal complexation of thiol-containing peptides from soy glycinin hydrolysates. Int J Mol Sci 16(4):8040–8058. https://doi.org/10.3390/ijms16048040
47. Hassan D, Bakhsh H, Khurram AM, Bhutto SA, Jalbani NS, Ghumro T, Solangi AR (2022) Fluorescent nanotechnology: an evolution in optical sensors. Curr Anal Chem 18(2):176–185. https://doi.org/10.2174/1573411017666201215121420
48. Sanjay BP, Sandeep S, Santhosh AS, Karthik CS, Varun DN, Kumara Swamy N, Mallu P, Nithin KS, Rajabathar JR, Muthusamy K (2022) Unprecedented 2D GNR-CoB nanocomposite for detection and degradation of malachite green—a computational prediction of degradation pathway and toxicity. Chemosphere 287:132153. https://doi.org/10.1016/j.chemosphere.2021.132153
49. Chelly M, Chelly S, Zribi R, Bouaziz-ketata H, Gdoura R, Lavanya N, Veerapandi G, Sekar C, Neri G (2021) Synthesis of silver and gold nanoparticles from rumex roseus plant extract and their application in electrochemical sensors. Nano 11(3):1–18. https://doi.org/10.3390/nano11030739
50. Verma SK, Das AK, Gantait S, Panwar Y, Kumar V, Brestic M (2022) Green synthesis of carbon-based nanomaterials and their applications in various sectors: a topical review. Carbon Lett 32(2):365–393. https://doi.org/10.1007/s42823-021-00294-7
51. Biranje A, Azmi N, Tiwari A, Chaskar A (2021) Quantum dots based fluorescent probe for the selective detection of heavy metal ions. J Fluoresc 31(5):1241–1250. https://doi.org/10.1007/s10895-021-02755-8
52. Aswathi VP, Meera S, Maria CGA, Nidhin M (2023) Green synthesis of nanoparticles from biodegradable waste extracts and their applications: a critical review. Nanotechnol Environ Eng 8(2):377–397. https://doi.org/10.1007/s41204-022-00276-8
53. Dhand C, Dwivedi N, Jun X, Ying A, Kumar N, Beuerman R, Lakshminarayanan R, Ramakrishna S (2018) Methods and strategies for the synthesis of diverse nanoparticles and their applications: a comprehensive overview. RSC Adv R Soc Chem 2:1–77
54. Chandra S, Kumar R, Singh R, Jain AK (2006) Coordination stability between metal/ligands interaction by modern spectroscopic studies: IR, electronic, EPR and cyclic voltammetry of cobalt(II) complexes with organic skeleton containing cyclic ligands. Spectrochim Acta Part A Mol Biomol Spectrosc 65(3–4):852–858. https://doi.org/10.1016/j.saa.2006.01.005
55. Vikraman AE, Thomas D, Cyriac ST, Kumar KG (2014) Kinetic and thermodynamic approach in the development of a voltammetric sensor for sunset yellow. J Electrochem Soc 161(14):B305–B311. https://doi.org/10.1149/2.0581414jes

56. Iravani S, Varma RS (2020) Sustainable synthesis of cobalt and cobalt oxide nanoparticles and their catalytic and biomedical applications. Green Chem 22(9):2643–2661. https://doi.org/10.1039/d0gc00885k
57. Ravikumar SB, Mallu TA, Subbareddy S, Shivamurthy SA, Neelalochana VD, Shantakumar KC, Rajabathar JR, Ataollahi N, Shadakshari S (2024) An enhanced non-enzymatic electrochemical sensor based on the Bi2S3-TiO2 nanocomposite with HNTs for the individual and simultaneous detection of 4-nitrophenol and nitrofurantoin in environmental samples. J Mater Chem B. https://doi.org/10.1039/d3tb03054g
58. Kim H, Na YJ, Song EJ, Kim KB, Bae JM, Kim C (2014) A single colorimetric sensor for multiple target ions: the simultaneous detection of Fe2+ and Cu2+ in aqueous media. RSC Adv 4(43):22463–22469. https://doi.org/10.1039/c4ra02776k
59. Polavarapu L, Pérez-Juste J, Xu QH, Liz-Marzán LM (2014) Optical sensing of biological, chemical and ionic species through aggregation of plasmonic nanoparticles. J Mater Chem C 2(36):7460–7476. https://doi.org/10.1039/c4tc01142b
60. Malapermal V, Botha I, Krishna SBN, Mbatha JN (2017) Enhancing antidiabetic and antimicrobial performance of ocimum basilicum, and ocimum sanctum (L.) using silver nanoparticles. Saudi. J Biol Sci 24(6):1294–1305. https://doi.org/10.1016/j.sjbs.2015.06.026
61. Kant T, Shrivas K, Tejwani A, Tandey K, Sharma A, Gupta S (2023) Progress in the design of portable colorimetric chemical sensing devices. Nanoscale 15(47):19016–19038. https://doi.org/10.1039/d3nr03803c
62. Kurup CP, Tlili C, Zakaria SNA, Ahmed MU (2021) Recent trends in design and development of nanomaterial-based aptasensors. Biointerface Res Appl Chem 11(6):14057–14077. https://doi.org/10.33263/BRIAC116.1405714077
63. Sargazi S, Fatima I, Hassan Kiani M, Mohammadzadeh V, Arshad R, Bilal M, Rahdar A, Díez-Pascual AM, Behzadmehr R (2022) Fluorescent-based nanosensors for selective detection of a wide range of biological macromolecules: a comprehensive review. Int J Biol Macromol 206(February):115–147. https://doi.org/10.1016/j.ijbiomac.2022.02.137
64. Cheng C, Xing M, Wu Q (2019) Green synthesis of fluorescent carbon dots/hydrogel nanocomposite with stable Fe3+ sensing capability. J Alloys Compd 790:221–227. https://doi.org/10.1016/j.jallcom.2019.03.053
65. Mandal T, Mishra SR, Singh V (2023) Comprehensive advances in the synthesis, fluorescence mechanism and multifunctional applications of red-emitting carbon nanomaterials. Nanoscale Adv 5(21):5717–5765. https://doi.org/10.1039/d3na00447c
66. Grandgirard J, Poinsot D, Krespi L, Nénon JP, Cortesero AM (2002) Costs of secondary parasitism in the facultative hyperparasitoid pachycrepoideus dubius: does host size matter? Entomol Exp Appl 103(3):239–248
67. Helim R, Zazoua A, Korri-Youssoufi H (2024) Sustainable biopolymer-based electrochemical sensors for trace heavy metal determination in water: a comprehensive review. Chem 12(12):10–28. https://doi.org/10.3390/chemosensors12120267
68. Uzunoğlu Doğruyol D (2023) Green synthesis and characterization of silver nanoparticles (AgNPs) and L-cysteine-capped AgNPs with foeniculum vulgare seed extract for colorimetric Hg2+ detection. Erzincan Üniversitesi Fen Bilim Enstitusu Derg 16(3):654–671. https://doi.org/10.18185/erzifbed.1352146
69. Beas-Bernuy LC, Cardenas-Riojas AA, Calderon-Zavaleta SL, Quiroz-Aguinaga U, La Rosa-Toro A, López EO, Asencios YJO, Baena-Moncada AM, Muedas-Taipe G (2023) Cd2+ detection by an electrochemical electrode based on MWCNT-orange peel activated carbon. ACS Omega 8(40):37341–37352. https://doi.org/10.1021/acsomega.3c05154
70. Mujahid M (2024) A simple green synthesis of carbon quantum dots from prunus armeniaca and their application as fluorescent probes for the selective and sensitive detection of cd 2+ metal ion. Mater Res Express. https://doi.org/10.1088/2053-1591/ad7921
71. Tadesse A, Hagos M, Ramadevi D, Basavaiah K, Belachew N (2020) Fluorescent-nitrogen-doped carbon quantum dots derived from citrus lemon juice: green synthesis, mercury(II)

ion sensing, and live cell imaging. ACS Omega 5(8):3889–3898. https://doi.org/10.1021/acsomega.9b03175
72. Ahmed SF, Mofijur M, Rafa N, Chowdhury AT, Chowdhury S, Nahrin M, Islam ABMS, Ong HC (2022) Green approaches in synthesising nanomaterials for environmental nanobioremediation: technological advancements, applications, benefits and challenges. Environ Res 204(March):111967. https://doi.org/10.1016/j.envres.2021.111967
73. Malik K, Kazmi A, Sultana T, Iqbal N, Bibi Y, Abbas M, Anjum I, Ali MM, Nasir M (2025) Heliyon review article a mechanistic overview on green assisted formulation of nanocomposites and their multifunctional role in biomedical applications. Heliyon 11(3):e41654. https://doi.org/10.1016/j.heliyon.2025.e41654
74. Abuzeid HM, Julien CM, Zhu L, Hashem AM (2023) Green synthesis of nanoparticles and their energy storage, environmental, and biomedical applications. Crystals 13(11):1576. https://doi.org/10.3390/cryst13111576
75. Salem MA, De Souza LP, Serag A, Fernie AR, Farag MA, Ezzat SM, Alseekh S (2020) Metabolomics in the context of plant natural products research: from sample preparation to metabolite analysis. Meta 10(1):1–30. https://doi.org/10.3390/metabo10010037
76. Barek J (2021) How to improve the performance of electrochemical sensors via minimization of electrode passivation. Chem 9(1):1–15. https://doi.org/10.3390/chemosensors9010012
77. Wuthnow M, Stafford M, Shih J (2009) IMS: a new model for blending applications

Chapter 6
Challenges and Future Directions in Heavy Metal Ion Sensing

H. M. Deepa, Mir Waqas Alam, Nagaraj Basavegowda, and Kiflom Gebremedhn Kelele

Abstract Heavy metal ions, such as lead (Pb^{2+}), cadmium (Cd^{2+}), mercury (Hg^{2+}), and arsenic (As^{3+}), pose significant environmental and health risks due to their toxicity, persistence, and bio-accumulative properties. Their presence in water, soil, air, and biological systems can lead to severe health issues, including neurological disorders, carcinogenic effects, and developmental abnormalities. Consequently, the accurate detection and quantification of heavy metal ions have become critical for environmental monitoring, public health safety, industrial processes, and regulatory compliance.

Traditional analytical techniques, such as atomic absorption spectroscopy (AAS), inductively coupled plasma mass spectrometry (ICP-MS), and electrochemical methods, offer high sensitivity and precision. However, these methods often require complex sample preparation, expensive equipment, and skilled personnel. This has driven the development of novel sensing technologies that are more cost-effective, portable, and capable of real-time monitoring.

Despite significant advancements, challenges remain in the field of heavy metal ion sensing. Issues related to selectivity, sensitivity, stability, and interference from complex matrices continue to hinder the development of robust and reliable sensors. Additionally, the need for rapid, on-site detection methods with minimal environmental impact has spurred ongoing research into innovative materials and sensor designs.

H. M. Deepa (✉)
Department of Studies in Chemistry, Davangere University, Davangere, Karnataka, India

M. W. Alam (✉)
Department of Physics, College of Science, King Faisal University, Al-Ahsa, Saudi Arabia
e-mail: wmir@kfu.edu.sa

N. Basavegowda
Department of Biotechnology, Yeungnam University, Gyeongsan, Korea

K. G. Kelele
Department of Chemistry, College of Natural and Computational Science, Woldia University, Woldia, Ethiopia

R. Naik et al. (eds.), *Biocompatible Nanomaterials for Smart Sensors*, Nanostructure Science and Technology,
https://doi.org/10.1007/978-3-032-01745-1_6

This chapter explores the current challenges in heavy metal ion sensing and outlines future directions aimed at overcoming these limitations. Advances in nanotechnology, biosensor integration, multifunctional sensors, and data-driven approaches are paving the way for more efficient, sustainable, and versatile sensing solutions.

Keywords Heavy metals · Cadmium · Lead · Mercury · Arsenic

6.1 Introduction

Urbanization is an ever-accelerating global trend that necessitates a thorough understanding of the relationships between human activities, the resulting emissions of pollutants, and the urban ecosystem [1]. Among the various contaminants studied, metals are particularly significant due to their widespread presence in urban settings, where they can accumulate rapidly and reach harmful concentrations. The situation is exacerbated by factors such as extensive infrastructure development, rising vehicle numbers, and shifts in lifestyle, particularly in developing nations, which increase both point and diffuse sources of metal emissions. Heavy metals are especially concerning due to their prevalence, toxicity, and persistence in the environment. The accumulation of these metals in urban soils, sediments, road dust, and water bodies poses a serious risk, as they can act as chemical time bombs, potentially contaminating groundwater when triggered by environmental changes [2–4]. The primary focus of heavy metal research is to identify the likely sources of pollution in urban areas and assess their implications for human health and the urban ecosystem. Paradoxically, the advancement of human society often coincides with increased metal pollution, making metals a focal point of environmental research over the past two decades [5–8]. In light of the vast amount of available information, a comprehensive review is essential to guide researchers toward contemporary research objectives. This chapter aims to outline the issue of heavy metal contamination in urban environments, emphasizing migration pathways and the significance of metal speciation and bioavailability. Additionally, it discusses various assessment tools for measuring metal levels in different environmental compartments, along with their limitations, and identifies potential future challenges in heavy metal research.

The rise in human activities has resulted in the discharge of numerous hazardous substances into water bodies, jeopardizing aquatic ecosystems and the surrounding environment. Among these pollutants, heavy metal ions pose the most significant threat due to their high toxicity, inability to degrade, and propensity to bioaccumulate and biomagnified through the food chain. Their presence in aquatic environments can adversely affect living organisms, both directly and indirectly [9, 10]. Furthermore, in terrestrial ecosystems, heavy metal ions are equally harmful to flora and fauna, as they can be absorbed by plants and subsequently enter the food chain, impacting animals and humans. In soil environments, the sorption of heavy metals by soil components plays a crucial role in mitigating the mobility of these pollutants, thereby protecting water sources, crops, and ultimately human and animal health. The consequences of water pollution have prompted researchers to devise effective methods for detecting and quantifying heavy metal contaminants [11]. Various techniques have been employed for this purpose, which can be categorized into three primary groups: spectroscopic detection methods, electrochemical detection methods, and optical detection methods. Spectroscopic techniques for identifying heavy metal ions include atomic absorption spectroscopy, atomic emission spectroscopy, inductively coupled plasma mass spectrometry, and cold vapor atomic fluorescence spectrometry. While these methods are highly sensitive, they tend to be costly and necessitate complex pre-treatment procedures [12–16].

In contrast, electrochemical methods are more economical, time-efficient, user-friendly, dependable, and appropriate for field applications. These methods encompass techniques such as amperometry, voltammetry, potentiometry, impedance measurement, and coulometry for the detection of heavy metal ions. The simplicity of these electrochemical techniques allows for rapid analytical processes, significantly quicker than many spectroscopic methods. Among the various electrochemical approaches, anodic stripping voltammetry is particularly favored for analyzing trace levels of heavy metal ions due to its high selectivity, portability, affordability, rapid analysis, and exceptional sensitivity. Conversely, optical detection methods identify metal ions through traditional techniques such as absorption, reflection, or luminescence spectrometry. In these optical methods, selective chromogenic reagents and indicator dyes are commonly employed, reacting specifically with targeted metal ions. Furthermore, tools like optical fibers, integrated optics, and capillary devices enhance the detection capabilities for metal ions.

The removal of heavy metal ions from aqueous environments is a critical issue, in addition to their detection. Water contaminated with highly toxic heavy metals such as Hg^{2+}, Pb^{2+}, and Cd^{2+} poses significant risks to human health, necessitating the development of highly effective and targeted removal techniques. Various methods can be utilized for this purpose, including precipitation, flocculation, membrane separation, ion exchange, and evaporation [17–21]. Each technique presents its own set of advantages and drawbacks regarding efficiency, sensitivity, selectivity, and specificity. Among these options, adsorption is frequently preferred due to its cost-effectiveness, straightforward design, and excellent operational capabilities, particularly its high efficiency in removing contaminants from dilute solutions. In many cases, the surfaces of materials are modified to improve their effectiveness in metal ion uptake.

6.2 Detection

A metal ion detector is an instrument specifically engineered to identify metal ions in its environment, and it may also facilitate the quantification of these ions. Given the critical significance of metal ion-free water systems, it is imperative to devise a strategy that ensures the safety of water bodies for living organisms. Before the removal of toxic metal ions from water samples, it is essential to have a reliable technique that can detect their presence and provide a quantitative assessment of pollution levels. This information is crucial for selecting an appropriate removal method. Therefore, the detection process should be developed to be both time-efficient and cost-effective, while also being environmentally friendly. Additionally, the detection technique must possess sufficient sensitivity to accurately identify even trace amounts of metal ions. Although various methods exist for detecting heavy metal ions, a universal technique applicable to all types of ions is currently lacking.

6.3 Detection Methods

6.3.1 Spectroscopic Detection

The spectroscopic identification of heavy metal ions employs highly sensitive methodologies such as atomic absorption spectroscopy, inductively coupled plasma mass spectrometry [22, 23], X-ray fluorescence spectrometry [24], neutron activation analysis, and inductively coupled plasma-optical emission spectrometry [25–27]. These methods are adept at simultaneously measuring the concentrations of various heavy metal ions across a wide range of elements, achieving remarkably low detection limits. Nonetheless, these techniques tend to be costly and necessitate skilled personnel to operate the intricate equipment. The flame atomic absorption spectroscopic method has been utilized to quantify Cu, Pb, and Cd ions in water, with detection limits reported at 2, 3, and 0.2 μg dm^{-3} for Cu^{2+}, Pb^{2+}, and Cd^{2+} ions, respectively [28]. Additionally, high-resolution surface plasmon resonance spectroscopy, in conjunction with anodic stripping voltammetry, has been employed to detect lead, copper, and mercury ions at concentrations ranging from parts per million to subparts per billion [29]. Below, several key spectroscopic techniques for the detection of heavy metal ions are discussed.

6.3.2 Atomic Absorption Spectroscopy

Atomic absorption spectroscopy employs a specific wavelength to transition isolated atoms from their ground state to an excited state. The energy absorbed during this excitation is quantified, and this value is directly proportional to the concentration of atoms in the sample. A typical atomic absorption spectrometer, which includes a primary light source, an atomizer for generating gas-phase atoms or ions for analysis, a monochromator, a detector, and an electronic readout system. The determination of elements through atomic absorption spectroscopy can only occur in a gaseous medium where individual atoms or ions are sufficiently separated. To produce gas-phase analyte atoms or ions, methods such as plasmas, flames, and electrothermal atomizers are utilized. The process of converting the sample into gas-phase atoms and ions is referred to as atomization. A specific light source, typically a hollow cathode lamp or an electrode-less discharge lamp corresponding to the element being analyzed, is employed to excite the atoms or ions. Additionally, there are multi-element lamps available that allow for the analysis of several elements without the need for lamp replacement. Solid-state detectors or photomultiplier tubes are commonly used as detectors in atomic absorption spectroscopy. For mercury determination, a specialized technique known as flow injection mercury systems is implemented. Table 6.1 summarizes the spectroscopic detection of various metal ions using atomic absorption spectroscopy.

Detecting heavy metals using atomic absorption spectroscopy (AAS) presents several challenges that can affect the accuracy and reliability of the results. One major issue is matrix effects, where complex sample matrices like biological tissues, soil, or industrial waste interfere with the atomization process, leading to inaccurate readings. Spectral interferences, such as overlapping absorption lines and background absorption from sample components, can further complicate detection. Additionally, chemical interferences arise when certain ions form nonvolatile compounds or react with reagents, hindering efficient atomization. The technique also faces limitations in sensitivity, as detecting trace levels of heavy metals requires advanced methods like Graphite Furnace AAS. Instrumental challenges, including the stability of the light source and flame consistency (in Flame AAS), can cause fluctuations in measurements. Sample preparation adds another layer of complexity, with risks of contamination, incomplete digestion, and time-consuming processes. Moreover, achieving accurate calibration is difficult due to the need for matrix-matched standards and potential instrumental drift over time. Despite these challenges, advancements in background correction techniques, the use of modifiers, and improved calibration methods have enhanced the capability of AAS in heavy metal detection.

Table 6.1 The spectroscopic detection of various metal ions by atomic absorption spectroscopy

Heavy metals	Technique	Reference
Mn, Pb, Cr, Cd	Flame atomic absorption spectrometry	[30]
Pb, Sb, Al, As	Atomic absorption spectroscopy	[31]
Cd, Cu, Fe, Ni, Pb, Zn	Atomic absorption spectrometry	[32]

6.4 Electrochemical Method

Electrochemical methods are cost-effective, user-friendly, and dependable, featuring straightforward procedures for the analysis of contaminated samples. One significant advantage of these methods is their rapid analytical time when compared to other spectroscopic techniques. Nonetheless, electrochemical techniques do have limitations, such as lower sensitivity and higher limits of detection relative to spectroscopic and optical methods [33]. Additionally, these techniques may require design enhancements and modifications to optimize their performance in detecting heavy metal ions [34]. For instance, various electrochemical methods often need to be integrated with different biosensing electrodes to enhance their sensitivity and detection limits by altering the electrode material. The presence of heavy metal ions in water typically results in changes to electrical parameters, including current, voltage, electrochemical impedance, charge, and electroluminescence within the electrochemical system [35]. Based on the influence of these electrical signals due to heavy metal ions, the techniques can be classified into categories such as potentiometric, amperometric, voltammetric, coulometric, impedance measurement, and electro chemiluminescent methods. Several key electrochemical techniques are discussed below.

6.5 Potentiometry

This method relies on measuring the electromotive force (EMF) without the need to apply any electric current during the measurement process. Potentiometric techniques are primarily employed for the quantitative analysis of ions in solutions [23]. Additionally, they prove to be effective in detecting heavy metal ions due to several advantages, including low cost, rapid response time, high selectivity, and a wide range of response [36, 37]. However, this technique does have limitations, such as high detection thresholds and decreased sensitivity, prompting ongoing efforts to address these issues. The use of modified electrodes, particularly those incorporating carbon nanotubes and metal nanoparticles, has shown promise in enhancing sensitivity and lowering detection limits for heavy metal ions.

Detecting heavy metals using potentiometry presents several challenges that can impact the accuracy, sensitivity, and reliability of the measurements. One major challenge is the selectivity of the electrodes, as ion-selective electrodes (ISEs) may respond to interfering ions present in complex samples, leading to inaccurate readings. Matrix effects also pose a problem, especially in environmental or biological samples, where the presence of other ions can alter the electrode's response. Additionally, pH sensitivity is a concern, as fluctuations in pH can affect the ion activity and, consequently, the potential difference measured. The low concentration of heavy metals can make detection challenging, requiring highly sensitive electrodes and careful optimization of measurement conditions. Electrode fouling

Table 6.2 The spectroscopic detection of various metal ions by different electrodes**a**

Heavy metal	Electrode	Reference
Hg	Screen printing electrode	[38]
Cd	PVC-based polyaniline	[39]
Pb	Poly pyrrole-modified electrode	[40]

and drift over time can degrade performance, leading to inconsistent results. Furthermore, sample preparation can introduce errors; improper dilution or contamination during handling can affect the accuracy of the analysis. Lastly, achieving accurate calibration is difficult due to potential interference from co-existing ions and the need for matrix-matched standards. Despite these challenges, potentiometry remains a valuable technique for heavy metal detection, especially when combined with advanced electrode materials and proper quality control measures (Table 6.2).

6.6 Amperometry

Amperometry is a potentiostatic method that involves applying a constant potential difference between a working electrode and a reference electrode within a solution containing electroactive species. This technique measures the flow of minute currents resulting from the reduction of heavy metal ions. The current is recorded over time, and these experiments are referred to as amperometric techniques. This method allows for the detection of a specific metal ion among various electroactive species due to the fixed potential of the working electrode. The analyte of interest undergoes a faradic reaction at a specified polarity and potential magnitude. However, the limited surface area of the working electrode often results in an incomplete faradic reaction, with only a portion of the analyte reacting. To address this issue, various modifications have been made to the electrode surface. For instance, a highly sensitive screen-printed electrode modified with a nanostructured carbon black film has been employed for the detection of Hg^{2+}, achieving a detection limit of 5 nM [41, 42]. Additionally, amperometric biosensors utilizing biologically modified electrodes have also been developed for the detection of heavy metal ions. These modifications aim to enhance the sensitivity of electrodes for metal ion detection using the amperometric technique.

Detecting heavy metals using amperometry presents several challenges that can affect the accuracy, sensitivity, and reliability of the results. One major challenge is electrode fouling and passivation, where the electrode surface can become contaminated with metal deposits or other substances, leading to reduced sensitivity and inconsistent responses over time. Interference from other ions is another significant issue, as non-target species can also undergo electrochemical reactions, causing false positives or skewed readings. The technique also struggles with selectivity, especially in complex samples like biological fluids or industrial waste, where multiple electroactive species are present. Signal instability can occur due to fluctuating current responses, often influenced by changes in temperature, pH, or ionic strength of the solution. Additionally, the need for precise control of experimental conditions, such as applied potential and electrode positioning, adds complexity to the analysis. Sample preparation can also introduce errors, as processes like filtration or digestion might affect the concentration of heavy metals. Finally, calibration challenges arise because the response of the electrode may not be linear over a wide range of concentrations, requiring careful standardization for accurate quantification. Despite these challenges, amperometry remains a powerful technique for heavy metal detection due to its high sensitivity, rapid response, and potential for in-situ analysis.

6.7 Voltammetry

Among the various electrochemical methods, voltammetry stands out as a prevalent technique for detecting heavy metal ions. This method involves measuring the current at different applied potentials to generate a current–voltage curve. The popularity of voltammetry in heavy metal ion detection can be attributed to its high accuracy, low detection limits, and exceptional sensitivity. While there are several types of voltammetry, the fundamental approach—measuring current by altering the potential—remains consistent across all forms. Below, we outline some of the specific modes of voltammetry utilized for analyzing heavy metal ions.

6.8 Cyclic Voltammetry

In cyclic voltammetry, the applied potential is initially varied in one direction before being reversed. The resulting current during both the forward and reverse scans is recorded and plotted against the applied potential, yielding a cyclic voltammogram. This voltammogram allows for the extraction of key parameters, such as the cathodic and anodic peak potentials along with their corresponding currents. A modified glassy carbon electrode (GCE) featuring gold nanoparticles and thiol-functionalized reduced graphene oxide (rGO-SH/Au nanoparticles) was utilized for the selective detection of Hg^{2+} ions through cyclic voltammetry [43]. The sensor demonstrated a linear response for Hg^{2+} concentrations ranging from 1 to 10 μM in a phosphate buffer saline (PBS) solution, achieving a low detection limit of 0.2 μM. The cyclic voltammetric response of the GCE/rGO-SH/Au nanoparticles was compared to that of GCE/rGO/Au nanoparticles without thiol functionalization, revealing a significant alteration in the anodic peak potential. The increased anodic currents observed with the GCE/rGO-SH/Au nanoparticles electrode system may be attributed to a higher loading of gold nanoparticles, as the –SH groups on the rGO sheets exhibit a strong affinity for the gold nanoparticles.

6.9 Differential Pulse Voltammetry

Pulse voltammetry refers to a technique that employs voltage signals of varying shapes and amplitudes in voltammetric measurements. This method is further divided into categories such as normal pulse voltammetry, reverse pulse voltammetry, and differential pulse voltammetry, among others. Among these techniques, differential pulse voltammetry is particularly favored for its heightened sensitivity in detecting heavy metal ions. A study conducted [44] demonstrated the simultaneous detection of copper, lead, and cadmium using a carbon paste electrode modified with hexagonal mesoporous silica (HMS) and immobilized quercetin (HMS-Qu/CPE). Voltammetric responses obtained from quercetin-modified carbon paste electrodes (Qu/CPE) and quercetin/ionic liquid-modified carbon paste electrodes (QuIL/CPE), highlight a comparison of their responses to Cu^{2+}, Pb^{2+}, and Cd^{2+} ions. The peak currents for copper, lead, and cadmium at the HMS-Qu/CPE exhibited significant enhancement, likely attributed to the synergistic effects of HMS and quercetin.

Both differential pulse voltammetry (DPV) and cyclic voltammetry (CV) face several challenges in the detection of heavy metals, which can affect their accuracy and reliability. In DPV, interference from co-existing ions and electroactive species can cause overlapping signals, leading to false positives. Electrode fouling, passivation, and surface contamination reduce sensitivity over time, while matrix effects from complex sample compositions can alter electrochemical responses. Additionally, DPV struggles with detecting ultra-trace metal levels, especially under varying pH conditions. In the case of CV, challenges include poor reproducibility due to inconsistent electrode surface modification and sensitivity to changes in scan rate, which can affect peak current and potential. Electrode fouling and adsorption of metal ions on the electrode surface can lead to signal drift, while overlapping redox peaks can complicate the interpretation of results. Both techniques also face difficulties with calibration, especially in the presence of complex matrices, and temperature fluctuations that influence reaction kinetics. Despite these limitations, improvements in electrode materials, nanostructures, and analytical techniques continue to enhance the performance of DPV and CV in heavy metal sensing.

6.10 Conclusion

The detection of heavy metal ions is a critical component in ensuring the safety of water bodies and protecting living organisms from the toxic effects of environmental pollution. Various techniques have been developed, each offering distinct advantages and facing specific limitations. Spectroscopic methods, such as atomic absorption spectroscopy (AAS), provide high sensitivity and accuracy for quantifying metal ions, though they require expensive equipment, skilled personnel, and are prone to interferences from complex sample matrices. Electrochemical methods, including potentiometry, amperometry, and voltammetry, offer cost-effective, rapid, and user-friendly alternatives with potential for in situ analysis. Among these, voltammetric techniques, particularly cyclic and differential pulse voltammetry, demonstrate exceptional sensitivity, low detection limits, and the ability to detect multiple metals simultaneously. However, challenges persist across all methods, including issues related to selectivity, matrix effects, interference from co-existing ions, and the need for precise calibration. Advancements in electrode materials, sensor design, and background correction techniques continue to improve detection capabilities. Ultimately, the choice of detection method depends on factors such as the required sensitivity, environmental conditions, cost, and the specific types of metal ions present. Integrating multiple techniques or enhancing traditional methods with novel materials may offer the most comprehensive solutions for accurate, reliable, and environmentally friendly heavy metal ion detection.

Funding The Deanship of Scientific Research, Vice Presidency for Graduate Studies and Scientific Research, King Faisal University, Saudi Arabia. [Project Number: KFU252507].

References

1. Nagajyoti PC, Lee KD, Sreekanth TVM (2010) Heavy metals, occurrence and toxicity for plants: a review. Environ Chem Lett 8:199–216. https://doi.org/10.1007/s10311-010-0297-8
2. Ayangbenro AS, Babalola OO (2017) A new strategy for heavy metal polluted environments: a review of microbial biosorbents. Int J Environ Res Public Health 14:94. https://doi.org/10.3390/ijerph14010094
3. Ali H, Khan E (2018) Bioaccumulation of non-essential hazardous heavy metals and metalloids in freshwater fish. Risk to human health. Environ Chem Lett 16(3):903–917. https://doi.org/10.1007/s10311-018-0734-7
4. Kahlon SK, Sharma G, Julka JM, Kumar A, Sharma S, Stadler FJ (2018) Impact of heavy metals and nanoparticles on aquatic biota. Environ Chem Lett 16:919–946. https://doi.org/10.1007/s10311-018-0737-4
5. Dubey S, Shri M, Gupta A, Rani V, Chakrabarty D (2018) Toxicity and detoxification of heavy metals during plant growth and metabolism. Environ Chem Lett 16:1169–1192. https://doi.org/10.1007/s10311-018-0741-8
6. Mallampati SR, Mitoma Y, Okuda T, Sakita S, Kakeda M (2013) Total immobilization of soil heavy metals with nano-Fe/ca/CaO dispersion mixtures. Environ Chem Lett 11:119–125. https://doi.org/10.1007/s10311-012-0384-0
7. Zhu J, Fu Q, Qiu G, Liu Y, Hu H, Huang Q, Violante A (2019) Influence of low molecular weight anionic ligands on the sorption of heavy metals by soil constituents: a review. Environ Chem Lett 17:1271. https://doi.org/10.1007/s10311-019-00881-1
8. Lair GJ, Gerzabek MH, Haberhauer G (2007) Sorption of heavy metals on organic and inorganic soil constituents. Environ Chem Lett 5:23–27. https://doi.org/10.1007/s10311-006-0059-9
9. Lebedev A, Sinikova N, Nikolaeva S, Poliakova O, Khrushcheva M, Pozdnyakov S (2003) Metals and organic pollutants in snow surrounding an iron factory. Environ Chem Lett 1:107–112. https://doi.org/10.1007/s10311-002-0004-5
10. Harrington CF, Clough R, Drennan-Harris LR, Hill SJ, Tyson JF (2011) Atomic spectrometry update. Elemental speciation. Journal of Analytical Atomic Spectromet 26(8):1561–1595. https://doi.org/10.1039/C1JA90030G

11. Zhu C, Yang G, Li H, Du D, Lin Y (2015) Electrochemical sensors and biosensors based on nanomaterials and nanostructures. Anal Chem 87(1):230–249. https://doi.org/10.1021/ac5039863
12. Zhu Y, Hu J, Wang J (2012) Competitive adsorption of Pb (II), cu (II) and Zn (II) onto xanthate-modified magnetic chitosan. J Hazard Mater 221:155–161. https://doi.org/10.1016/j.jhazmat.2012.04.026
13. Zhang QX, Wen H, Peng D, Fu Q, Huang XJ (2015) Interesting interference evidences of electrochemical detection of Zn (II), cd (II) and Pb (II) on three different morphologies of MnO2 nanocrystals. J Electroanal Chem 739:89–96. https://doi.org/10.1016/j.jelechem.2014.12.023
14. Yao XZ, Guo Z, Yuan QH, Liu ZG, Liu JH, Huang XJ (2014) Exploiting differential electrochemical stripping behaviors of Fe3O4 nanocrystals toward heavy metal ions by crystal cutting. ACS Appl Mater Interfaces 6(15):12203–12213. https://doi.org/10.1021/am501617a
15. Liu ZG, Chen X, Liu JH, Huang XJ (2014) Robust electrochemical analysis of as (III) integrating with interference tests: a case study in groundwater. J Hazard Mater 278:66–74. https://doi.org/10.1016/j.jhazmat.2014.05.094
16. Xu RX, Yu XY, Gao C, Liu JH, Compton RG, Huang XJ (2013) Enhancing selectivity in stripping voltammetry by different adsorption behaviors: the use of nanostructured Mg–Al-layered double hydroxides to detect cd (II). Analyst 138(6):1812–1818. https://doi.org/10.1039/C3AN36271J
17. Pujol L, Evrard D, Groenen-Serrano K, Freyssinier M, Ruffien-Cizsak A, Gros P (2014) Electrochemical sensors and devices for heavy metals assay in water: the French groups' contribution. Front Chem 2:19. https://doi.org/10.1016/j.elecom.2006.07.041
18. Crini G, Lichtfouse E (2019) Advantages and disadvantages of techniques used for wastewater treatment. Environ Chem Lett 17:145–155. https://doi.org/10.1007/s10311-018-0785-9
19. Wang X, Deng W, Xie Y, Wang C (2013) Selective removal of mercury ions using a chitosan–poly (vinyl alcohol) hydrogel adsorbent with three-dimensional network structure. Chem Eng J 228:232–242. https://doi.org/10.1016/j.cej.2013.04.104
20. Elfeky SA, Mahmoud SE, Youssef AF (2017) Applications of CTAB modified magnetic nanoparticles for removal of chromium (VI) from contaminated water. J Adv Res 8(4):435–443. https://doi.org/10.1016/j.jare.2017.06.002
21. Ojemaye MO, Okoh OO, Okoh AI (2017) Surface modified magnetic nanoparticles as efficient adsorbents for heavy metal removal from wastewater: Progress and prospects. Mater Express 7(6):439–456. https://doi.org/10.1166/mex.2017.1401
22. Gong T, Liu J, Liu X, Liu J, Xiang J, Wu Y (2016) A sensitive and selective sensing platform based on CdTe QDs in the presence of L-cysteine for detection of silver, mercury and copper ions in water and various drinks. Food Chem 213:306–312. https://doi.org/10.1016/j.foodchem.2016.06.091
23. Aragay G, Merkoçi A (2012) Nanomaterials application in electrochemical detection of heavy metals. Electrochim Acta 84:49–61. https://doi.org/10.1016/j.electacta.2012.04.044
24. Wang H, Wu Z, Chen B, He M, Hu B (2015) Chip-based array magnetic solid phase microextraction on-line coupled with inductively coupled plasma mass spectrometry for the determination of trace heavy metals in cells. Analyst 140(16):5619–5626. https://doi.org/10.1039/C5AN00736D
25. Sitko R, Janik P, Zawisza B, Talik E, Margui E, Queralt I (2015) Green approach for ultratrace determination of divalent metal ions and arsenic species using total-reflection X-ray fluorescence spectrometry and mercapto-modified graphene oxide nanosheets as a novel adsorbent. Anal Chem 87(6):3535–3542. https://doi.org/10.1021/acs.analchem.5b00283
26. Losev VN, Buyko OV, Trofimchuk AK, Zuy ON (2015) Silica sequentially modified with polyhexamethylene guanidine and Arsenazo I for preconcentration and ICP–OES determination of metals in natural waters. Microchem J 123:84–89. https://doi.org/10.1016/j.microc.2015.05.022
27. Pöykiö R, Perämäki P (2003) Acid dissolution methods for heavy metals determination in pine needles. Environ Chem Lett 1:191–195. https://doi.org/10.1007/s10311-003-0038-3

28. Shirkhanloo H, Mousavi ZH, Rouhollahi A (2011) Preconcentration and determination of heavy metals in water, sediment and biological samples. J Serb Chem Soc 76(11):1583–1595. https://doi.org/10.2298/JSC101024139S
29. Wang S, Forzani ES, Tao N (2007) Detection of heavy metal ions in water by high-resolution surface plasmon resonance spectroscopy combined with anodic stripping voltammetry. Anal Chem 79(12):4427–4432. https://doi.org/10.1021/ac0621773
30. Tsade HK (2016) Atomic absorption spectroscopic determination of heavy metal concentrations in Kulufo River, Arbaminch, Gamo Gofa. Ethiopia J Environ Anal Chem 3(177):2. https://doi.org/10.4172/2380-2391.1000177
31. Tareen AK, Sultan IN, Parakulsuksatid P, Shafi M, Khan A, Khan MW, Hussain S (2014) Detection of heavy metals (Pb, Sb, Al, As) through atomic absorption spectroscopy from drinking water of District Pishin, Balochistan, Pakistan. Int J Curr Microbiol Appl Sci 3(1):299–308. ISSN: 2319–7706 Volume 3 Number 1 (2014) pp.299–308
32. Kojuncu Ý, Bundalevska JM, Ay Ü, Čundeva K, Stafilov T, Akçin G (2004) Atomic absorption spectrometry determination of cd, cu, Fe, Ni, Pb, Zn, and TI traces in seawater following flotation separation. Sep Sci Technol 39(11):2751–2765. https://doi.org/10.1081/SS-200026751
33. Pujol L, Evrard D, Groenen-Serrano K, Freyssinier M, Ruffien-Cizsak A, Gros P (2014) Electrochemical sensors and devices for heavy metals assay in water: the French groups' contribution. Front Chem 2:19. https://doi.org/10.3389/fchem.2014.00019
34. Bansod B, Kumar T, Thakur R, Rana S, Singh I (2017) A review on various electrochemical techniques for heavy metal ions detection with different sensing platforms. Biosens Bioelectron 94:443–455. https://doi.org/10.1016/j.bios.2017.03.031
35. Cui L, Wu J, Ju H (2015) Electrochemical sensing of heavy metal ions with inorganic, organic and bio-materials. Biosens Bioelectron 63:276–286. https://doi.org/10.1016/j.bios.2014.07.052
36. Düzgün A, Zelada-Guillén GA, Crespo GA, Macho S, Riu J, Rius FX (2011) Nanostructured materials in potentiometry. Anal Bioanal Chem 399:171–181. https://doi.org/10.1007/s00216-010-3974-3
37. Bakker E, Pretsch E (2008) Nanoscale potentiometry. TrAC Trends Anal Chem 27(7):612–618. https://doi.org/10.1016/j.trac.2008.04.007
38. Aglan RF, Saleh HM, Mohamed GG (2018) Potentiometric determination of mercury (II) ion in various real samples using novel modified screen-printed electrode. Appl Water Sci 8:1–11. https://doi.org/10.1007/s13201-018-0781-z
39. Naushad M, Vasudevan S, Sharma G, Kumar A, Alothman ZA (2016) Adsorption kinetics, isotherms, and thermodynamic studies for Hg2+ adsorption from aqueous medium using alizarin red-S-loaded amberlite IRA-400 resin. Desalin Water Treat 57(39):18551–18559. https://doi.org/10.1080/19443994.2015.1090914
40. Mazloum-Ardakani M, Amini MK, Dehghan M, Kordi E, Sheikh-Mohseni MA (2012) Nanomolar determination of Pb (II) ions by selective templated electrode. J Serb Chem Soc 77(7):899–910. https://doi.org/10.2298/JSC110830210M
41. Arduini F, Majorani C, Amine A, Moscone D, Palleschi G (2011) Hg2+ detection by measuring thiol groups with a highly sensitive screen-printed electrode modified with a nanostructured carbon black film. Electrochim Acta 56(11):4209–4215. https://doi.org/10.1016/j.electacta.2011.01.094
42. Mohammadi H, Amine A, Cosnier S, Mousty C (2005) Mercury–enzyme inhibition assays with an amperometric sucrose biosensor based on a trienzymatic-clay matrix. Anal Chim Acta 543(1–2):143–149. https://doi.org/10.1016/j.aca.2005.04.014
43. Devi NR, Sasidharan M, Sundramoorthy AK (2018) Gold nanoparticles-thiol-functionalized reduced graphene oxide coated electrochemical sensor system for selective detection of mercury ion. J Electrochem Soc 165(8):B3046–B3053. https://doi.org/10.1149/2.0081808jes
44. Xia F, Zhang X, Zhou C, Sun D, Dong Y, Liu Z (2010) Simultaneous determination of copper, lead, and cadmium at hexagonal mesoporous silica immobilized quercetin modified carbon paste electrode. J Autom Methods Manag Chem 2010:824197. https://doi.org/10.1155/2010/824197

Chapter 7
Nanomaterials for Uric Acid, Urea, and Creatinine Sensing

Dinakaran Thirumalai and Aoife Morrin

Abstract The detection of uric acid, urea, and creatinine—key biomarkers for renal and metabolic disorders—has significantly advanced with the development of nanomaterial-based biosensors. This chapter presents a comprehensive review of recent progress in electrochemical and optical sensing platforms that leverage the unique physicochemical properties of nanomaterials such as metallic nanoparticles, metal oxide nanostructures, carbon-based materials, and polymeric nanocomposites. These materials enhance biosensor performance by increasing electron transfer efficiency, catalytic activity, and surface functionalization, thereby improving sensitivity, selectivity, and stability. Enzyme-based and non-enzymatic detection strategies are discussed, alongside emerging applications in flexible, wearable, and smartphone-integrated devices. Special emphasis is placed on molecularly imprinted polymers, metal-organic frameworks, and hybrid nanocomposites as next-generation platforms for point-of-care diagnostics. The chapter also explores challenges related to biocompatibility, scalability, and regulatory compliance, and outlines future directions including the use of MXenes, AI integration, and biodegradable sensors. These innovations collectively position nanomaterial-enabled biosensing as a transformative approach for real-time, personalized healthcare monitoring.

Keywords Uric acid · Urea · Creatinine

D. Thirumalai (✉) · A. Morrin
Insight Research Ireland Centre for Data Analytics, School of Chemical Sciences, Dublin City University, Dublin, Ireland
e-mail: dinakaran.thirumalai@dcu.ie

R. Naik et al. (eds.), *Biocompatible Nanomaterials for Smart Sensors*, Nanostructure Science and Technology,
https://doi.org/10.1007/978-3-032-01745-1_7

7.1 Introduction

Nanomaterials are particles or structures that measure within the nanometer range (1–100 nm) in at least one dimension, making them significantly smaller than typical bulk materials. Their extremely small size gives rise to unique chemical and physical properties that differ significantly from those found in bulk materials. These differences stem from phenomena specific to the nanoscale dimensions, such as increased surface-to-volume ratios, quantum confinement effects, and enhanced reactivity. Controlling and manipulating these characteristics enables the development of novel products, innovative technologies, and advanced scientific instruments. The recent development of sophisticated synthesis methods and precise manipulation tools has greatly accelerated research interest in nanomaterials. Numerous advanced fabrication methods now exist, facilitating detailed exploration of how size influences the electrical, optical, and magnetic properties of nanostructures.

A biosensor is an integrated device designed to analyze information obtained from a biological component connected to a signal transduction element. Biosensors can detect various substances, including gases, heavy metal ions, carbohydrates, amino acids, and disease-associated molecules. Early biosensors were only suitable for basic detection tasks; however, ongoing technological improvements and refinements have enabled modern biosensors to operate with higher sensitivity, enhanced stability, lower cost, and simpler use. In this development, nanomaterials play a critical role. They are primarily integrated into biosensors to boost sensitivity and amplify the detection signals. Additionally, nanomaterials can be functionalized to improve other biosensor attributes, including robustness, stability, linear response, repeatability, biocompatibility, resolution, and response speed. Specifically, nanomaterials enhance biosensors by significantly increasing effective surface areas, facilitating biomolecule immobilization, promoting electrochemical and optical reactions while decreasing activation energies, accelerating electron transport processes, separating biomolecule complexes, concentrating analytes before detection, and serving as labels for biological recognition elements.

Uric acid (UA) is the end product of purine metabolism and typically appears as white crystals. It has limited solubility in water and human body fluids, primarily existing as urate salts. The kidneys eliminate UA from the body via urine. In many organisms, UA is oxidized into allantoin by the enzyme uricase, producing hydrogen peroxide as a byproduct. However, humans lack uricase, preventing this oxidation and causing UA to accumulate when levels become elevated. Excess UA deposits in joints, tendons, and surrounding tissues—especially in extremities—leading to gout, a type of arthritis. Elevated UA concentrations are also associated with kidney stones, hypertension, metabolic syndrome, chronic kidney disease, and cardiovascular diseases. Conversely, lower-than-normal UA levels have been connected to multiple sclerosis and certain genetic disorders [1].

Urea is an important marker of nitrogen metabolism, with its concentration reflecting the body's metabolic condition and kidney function. The physiological concentration of urea in human serum typically ranges from 15 to 40 mg/dL

(2.5–7.5 mM/L). Deviations from this normal range are associated with various pathological conditions. Elevated urea levels (hyperuremia) can lead to or indicate disorders such as renal dysfunction, including kidney failure, urinary tract obstruction, dehydration, gastrointestinal bleeding, shock, burns, ulcers, and certain malignancies. Conversely, abnormally low urea concentrations may be indicative of hepatic failure, nephrotic syndrome, or cachexia [2].

Creatinine is one of the most widely used indicators of kidney function. It is a waste product formed from the metabolism of creatine phosphate in muscles and from dietary protein breakdown. In the human body, creatinine comes from two main sources: endogenous creatinine, produced by muscle tissue; and exogenous creatinine, derived from the metabolism of meat in the diet. Under normal conditions, creatinine is produced at a relatively constant rate and is almost entirely filtered by the kidneys and excreted in urine. As a result, serum creatinine levels typically remain stable, ranging between 44 and 106 μmol/L. However, when kidney function is impaired and the body's ability to eliminate creatinine decreases, its concentration in the blood rises above the normal physiological range [3]. Therefore, detecting and monitoring UA, urea, and creatinine have become a significant focus in healthcare and medical diagnostics research.

The determination of UA, urea, and creatinine can be achieved through a range of analytical methodologies, encompassing both conventional and advanced techniques. Conventional approaches include chromatography and enzyme-based assays, which are known for their relatively high specificity. However, these approaches may exhibit limitations in terms of sensitivity, reproducibility, and analytical reliability under certain conditions. In recent years, the development of nanotechnology-based detection platforms—such as nanomaterial-assisted assays, electrochemical biosensors, optical sensors, and chemiluminescence-based systems—has significantly advanced the field. These modern techniques offer superior analytical performance, including enhanced sensitivity, improved signal-to-noise ratios, and greater reproducibility, thereby addressing key limitations inherent to conventional methodologies [4]. As a result, nanomaterial-based sensing devices are increasingly being recognized for their potential in accurate and early detection of these critical biomarkers.

7.2 Types of Nanomaterials for Biosensing

Advances in the synthesis and functionalization of nanomaterials have facilitated the design and fabrication of multifunctional biosensors with improved sensitivity, specificity, and stability. Also, the biocompatibility of nanomaterials have an impact on their application in biosensing. Consequently, nanomaterials have become integral components in modern biosensing platforms [5]. A diverse range of nanomaterials has been introduced to enhance the performance of biosensors, including metallic nanoparticles, metal oxide nanoparticles, carbon-based nanomaterials, and polymeric nanomaterials.

7.2.1 Metallic Nanoparticles

Metallic nanoparticles (NPs), particularly those composed of noble metals such as gold (Au), silver (Ag), platinum (Pt), and palladium (Pd), have garnered significant attention in biosensing applications due to their unique optical, electronic, and catalytic properties [6]. These nanoparticles play a pivotal role in enhancing electron transfer kinetics and catalytic activity, thereby improving the sensitivity and selectivity of electrochemical biosensors for analytes like UA, urea, and creatinine. Also, Au and AgNPs are extensively utilized in plasmonic sensing platforms.

Electrochemical sensing platforms also benefit significantly from the incorporation of AuNPs and AgNPs. These NPs offer a high surface-to-volume ratio, superior electrical conductivity, and excellent electrocatalytic activity, which contribute to enhanced electron transfer kinetics and signal amplification. AuNPs, for instance, provide a stable and biocompatible surface for enzyme immobilization and facilitate redox reactions, thereby improving the overall performance of amperometric and voltammetric biosensors. AgNPs, on the other hand, have been shown to increase current response due to their excellent charge transfer capabilities and surface reactivity.

Recent advancements have demonstrated the efficacy of AuNPs and AgNPs in the sensitive detection of UA, urea, and creatinine. For instance, a study developed a non-enzymatic sensor for UA detection based on AuNPs electrodeposited on laser-induced graphene electrodes, achieving high sensitivity and selectivity [7]. Another research utilized nanoporous Au for the sensitive detection of UA and creatinine in human urine, leveraging the unique electrochemical oxidation properties of nanoporous Au toward UA [8].

In the field of green chemistry, biosynthesized AgNPs anchored on a carbon material derived from maple leaves have been employed to develop a non-enzymatic biosensor for creatinine sensing, showcasing the potential of eco-friendly materials in biosensor fabrication [9].

A study developed an electrochemical biosensing platform using palladium nanoparticle-decorated copper-hemin metal-organic frameworks for creatinine detection. The PdNPs provided superior catalytic activity, resulting in enhanced sensor sensitivity [10].

In plasmonic sensing, AuNPs and AgNPs exhibit localized surface plasmon resonance (LSPR), a phenomenon characterized by the coherent oscillation of conduction electrons on the NP surface when excited by incident light at specific wavelengths. This results in significant amplification of the local electromagnetic field, enhancing the sensitivity of biosensors to changes in the surrounding dielectric environment. Functionalization of these NPs with selective biorecognition elements (e.g., enzymes or aptamers) allows for the label-free and real-time detection of target analytes, with shifts in the LSPR absorption peak serving as the analytical signal.

7.2.2 Metal Oxide Nanoparticles

Metal oxide nanoparticles (MONPs) have become integral components in biosensing technologies due to their unique chemical and physical properties, such as high surface area, catalytic activity, electron mobility, and ease of functionalization [11]. Their robust chemical stability and biocompatibility make them ideal candidates for developing sensors capable of detecting small biomolecules like UA, urea, and creatinine.

Commonly studied MONPs include zinc oxide (ZnO), titanium dioxide (TiO_2), iron oxide (Fe_3O_4), cerium oxide (CeO_2), cobalt oxide (CoO), nickel oxide (NiO), and copper oxide (CuO). These materials offer several advantages:

- Enhanced electron transfer rates.
- High loading capacity for biomolecules.
- Excellent catalytic properties.
- Stability under physiological conditions.
- Biocompatibility.

These attributes significantly improve the sensitivity, selectivity, and detection limits of biosensors [12].

For UA sensing, MONPs often serve as electrocatalysts, promoting direct oxidation of uric acid at lower potentials. ZnO nanoparticles, in particular, are known for their strong interaction with UA molecules, leading to amplified current signals [13]. For instance, ZnO nanorod-modified electrodes have demonstrated excellent selectivity and sensitivity for UA in complex biological fluids, minimizing interference from common substances like ascorbic acid and dopamine [14].

Urea sensors typically involve immobilized urease enzymes, where metal oxides act as supportive platforms. TiO_2 nanoparticles offer high surface area and favorable biocompatibility, facilitating effective enzyme immobilization and maintaining enzyme activity. Upon urea hydrolysis, the generated ammonium ions alter the local pH, which can be detected electrochemically. A TiO_2-based biosensor has shown excellent analytical performance with low detection limits and good operational stability [15]. CuO nanoparticles exhibit significant electrocatalytic activity, making them effective in non-enzymatic biosensing platforms. Recent studies have demonstrated the use of CuO/Co_3O_4 nanocomposites in screen-printed electrodes for the direct detection of urea, achieving high sensitivity and stability [16].

Creatinine sensing is vital for renal health assessment. Metal oxide nanoparticles such as TiO_2 have been utilized to develop non-enzymatic creatinine sensors. TiO_2 nanoparticles exhibit intrinsic redox activity, allowing direct electron transfer between creatinine molecules and the electrode surface [17]. Such sensors offer advantages including simplicity, lower cost, and better storage stability compared to enzyme-based systems [18].

7.2.3 Carbon Nanomaterials

Carbon nanomaterials (CNMs), including graphene, carbon nanotubes (CNTs), graphene oxide (GO), reduced graphene oxide (rGO), and carbon quantum dots (CQDs), have garnered significant attention in biosensing applications due to their exceptional electrical conductivity, large surface area, and ease of functionalization. These properties make CNMs ideal candidates for developing sensitive and selective biosensors for detecting renal biomarkers such as UA, urea, and creatinine.

Graphene and its derivatives, particularly GO and rGO, offer a two-dimensional structure with a high surface-to-volume ratio, facilitating the immobilization of biomolecules and enhancing electron transfer. Their integration into biosensors has led to improved sensitivity and selectivity in detecting analytes like UA and creatinine. For instance, a study developed a ternary nanocomposite composed of polymer poly(3,4-ethylenedioxythiophene)-poly(styrene sulfonate), AuNPs, and electrochemically rGO on a glassy carbon electrode for the simultaneous detection of dopamine and UA in human urine, demonstrating excellent ion selectivity and electrocatalytic activity [19].

Carbon nanotubes (CNTs), including single-walled (SWCNTs) and multiwalled (MWCNTs) variants, exhibit remarkable electrical conductivity and mechanical strength. Their incorporation into biosensors has been explored for various analytes, contributing to improved detection capabilities. Recent studies have demonstrated the use of CNT-based biosensors for the direct detection of urea, achieving high sensitivity and stability. For example, a machine learning-assisted, flexible, electrochemical, non-enzymatic biosensor was developed using a novel MWCNT-ZnO nanocomposite functionalized with CuO-micro-flowers, enhancing sensitivity and operational performance for precise urea concentration detection [20]. β-lead dioxide/single-walled carbon nanotube (β-PbO_2/SWCNT) is used as highly sensitive potentiometric sensors for creatinine detection [21].

Carbon quantum dots (CQDs) are zero-dimensional carbon nanomaterials known for their excellent photoluminescence properties and biocompatibility. Their integration into biosensors has been explored for various analytes, contributing to improved detection capabilities. Recent studies have demonstrated the use of CQD-based biosensors for the direct detection of creatinine, achieving high sensitivity and stability. For instance, a study reported the development of a novel, eco-friendly non-enzymatic biosensor for creatinine determination in synthetic urine, utilizing biosynthesized silver nanoparticles anchored on a carbon material derived from natural sources [9].

7.2.4 Polymeric Nanomaterials

Polymeric nanomaterials have garnered significant attention in biosensing applications due to their tuneable physicochemical properties, biocompatibility, and versatility in functionalization.

Conducting polymers like polyaniline (PANI), polypyrrole (PPy), and poly(3,4-ethylenedioxythiophene) (PEDOT) are widely utilized in biosensor fabrication owing to their excellent electrical conductivity and ease of synthesis. Their incorporation into biosensors has led to enhanced electron transfer rates and improved sensitivity in detecting analytes like creatinine. For instance, integrating urease into PEDOT-based organic electrochemical transistors (OECTs) enabled urea sensing through transistor response changes resulting from local pH variations due to enzymatic catalysis [22].

Molecularly imprinted polymers (MIPs) are synthetic polymers with specific recognition sites tailored for target molecules, offering high selectivity and stability. Their application in biosensors for renal biomarkers has shown promising results. A recent study reported the development of an electrochemical biosensor employing a MIP for creatinine detection, achieving a low detection limit and excellent reproducibility [23]. Another study has demonstrated that UA molecularly imprinted polyaniline (PAn)-coated Cu-zeolitic imidazolate framework (Cu-ZIF) nanoparticles—termed "polynanozymes"—significantly enhance nanozyme catalytic activity [24].

The integration of polymers with nanomaterials such as carbon nanotubes, graphene, and metal nanoparticles forms nanocomposites that synergistically enhance biosensor performance. These composites exhibit improved mechanical strength, electrical conductivity, and biocompatibility. For example, a novel electrochemical sensor utilizing a layer-by-layer architecture of poly(amidoamine) dendrimer (PAMAM D-G0.5), multiwalled carbon nanotube-silver nanoparticles (MWCNT-AgNP), and poly(neutral red) (poly[NR]) was developed for rapid and direct UA detection [25].

Hydrogels and nanogels are three-dimensional polymeric networks capable of retaining large amounts of water, providing a conducive environment for biomolecule immobilization [26]. Their responsive nature to environmental stimuli makes them suitable for biosensing applications [27, 28]. Recent advancements include a flexible, wearable contact lens sensor based on a nickel-cobalt metal-organic framework (Ni-Co-MOF) hydrogel that was developed for noninvasive urea monitoring in tear samples. The sensor demonstrated excellent selectivity against common interferents such as UA, ascorbic acid, glucose, and various ions. Enhanced electrocatalytic activity was attributed to Ni^{2+} oxidation and Co^{2+} complex formation. The nontoxic, binder-free hydrogel offers promising applications for wearable biosensors, point-of-care diagnostics, and personalized health monitoring [29].

7.3 Mechanisms of Nanomaterial-Based Biosensing

The integration of nanomaterials into sensing platforms has revolutionized the detection of critical biomarkers such as UA, urea, and creatinine. Their unique physicochemical properties significantly improve sensitivity, selectivity, and detection limits. This section discusses the key mechanisms by which nanomaterials enhance sensor performance, categorized under electrochemical, optical, enzyme-based, and non-enzymatic approaches.

7.3.1 *Electrochemical Sensors*

7.3.1.1 Nanomaterials Enhancing Electron Transfer

Nanomaterials like metal NPs (Au, Ag), CNTs, graphene, and metal oxides (ZnO, TiO_2) offer high surface areas and excellent electrical conductivity. These properties facilitate faster electron transfer between the analyte and the electrode surface, leading to amplified current responses and lower detection limits. Surface functionalization with nanomaterials creates more active sites, increasing the interaction between target molecules and the electrode.

Key Points
- Reduced electron-transfer resistance improves signal response.
- Increased electrode surface area leads to more analyte adsorption.
- Better conductivity and catalytic activity amplify electrochemical signals.

7.3.1.2 Examples of Electrochemical Techniques

Several electrochemical techniques are employed in combination with nanomaterials for biomarker sensing:

Amperometry: Measures current at a constant voltage, ideal for real-time monitoring of enzymatic reactions catalyzing UA or urea.

Voltammetry: Techniques such as cyclic voltammetry (CV) and differential pulse voltammetry (DPV) detect oxidation/reduction peaks, improving detection sensitivity for UA and creatinine.

Electrochemical Impedance Spectroscopy (EIS): Provides information on surface modifications and charge transfer processes, useful for characterizing nanomaterial-modified electrodes.

7.3.2 *Optical Sensors*

7.3.2.1 Detection Mechanisms: Fluorescence, Colorimetry, and Plasmonics

Fluorescence-Based Detection: Quantum dots and carbon dots are widely used. The fluorescence intensity changes upon interaction with the target analyte, allowing sensitive detection of urea and creatinine.

Colorimetric Detection: Au and AgNPs exhibit size and aggregation-dependent color changes, enabling naked-eye detection of biomarkers.

Plasmonic Detection: LSPR in noble metal NPs enhances light scattering, allowing ultrasensitive optical biosensing.

7.3.2.2 Use of Quantum Dots and Noble Metal Nanoparticles

Quantum dots provide stable and bright fluorescence ideal for multiplexed detection. Noble metal NPs (like Au and Ag) are crucial for plasmonic sensors because of their strong surface plasmon resonance, leading to highly sensitive assays.

7.3.3 Enzyme-Based Vs. Non-enzyme-Based Sensors

7.3.3.1 Enzyme Immobilization on Nanomaterials

In enzyme-based sensors, biomolecules such as uricase, urease, and creatinine deiminase are immobilized onto nanomaterial surfaces to improve the biosensor's analytical performance. This immobilization not only stabilizes the enzyme structure, prolonging its functional lifespan, but also enhances electron transfer between the enzyme's active sites and the electrode surface, thereby improving signal transduction.

Additionally, nanomaterials such as graphene, metal oxides, and polymeric supports offer high surface area and biocompatibility, facilitating effective enzyme loading and preserving their catalytic activity. As a result, these sensors achieve high sensitivity, rapid response times, and improved reproducibility, making them highly suitable for clinical and point-of-care applications.

7.3.3.2 Non-enzymatic Electrochemical Oxidation and Catalytic Detection

Non-enzymatic sensors address many of the limitations associated with enzyme-based systems, particularly the issues of enzyme instability, limited shelf life, and sensitivity to environmental conditions such as pH and temperature. In these systems, nanomaterials serve as direct electrocatalysts for the oxidation of target analytes like UA, urea, and creatinine, eliminating the need for biological recognition elements. Metal oxides such as CuO, ZnO, and NiO, as well as noble metals like Au and Pt, exhibit excellent catalytic properties, enabling efficient electron transfer and redox reactions at the electrode surface. These materials not only facilitate high sensitivity and rapid detection but also enhance the robustness, chemical stability, and reusability of the sensors.

Additionally, non-enzymatic sensors are capable of functioning under a wider range of operating conditions, making them more suitable for long-term monitoring, harsh environments, and integration into wearable and portable diagnostic devices.

7.4 Nanomaterials for Individual Biomarkers

The detection of UA, urea, and creatinine as key clinical biomarkers has been significantly enhanced through the application of nanomaterials. Their unique properties—such as large surface area, tunable conductivity, and catalytic activity—enable both electrochemical and optical biosensing approaches. This section discusses nanomaterial strategies tailored to each biomarker.

7.4.1 Uric Acid Sensing

7.4.1.1 Nanomaterial-Based Strategies for Electrochemical and Optical Detection

UA is a critical biomarker for diagnosing disorders such as gout, kidney diseases, and cardiovascular problems. Nanomaterials have revolutionized UA sensing by offering improved sensitivity, selectivity, and stability.

- **Electrochemical Approaches:** Nanomaterials such as GO, ZnO nanoparticles, and CNTs enhance electron transfer and reduce detection limits in voltammetric and amperometric assays.
- **Optical Approaches:** AuNPs and quantum dots provide colorimetric and fluorescence-based detection of UA. Changes in fluorescence intensity or color shift serve as reliable indicators of UA concentration.

7.4.1.2 Examples: Gold Nanoparticle and Graphene-Based Uric Acid Sensors

- **Gold Nanoparticles (AuNPs):** Functionalized AuNPs have been employed for colorimetric detection of UA, where aggregation causes a visible color change, enabling rapid and simple assays [30, 31].
- **Graphene-Based Sensors:** Graphene and its derivatives (e.g., GO, rGO) offer high surface areas and excellent conductivity, facilitating low-potential oxidation of UA with superior sensitivity [32].

7.4.2 Urea Sensing

7.4.2.1 Role of Urease-Based Biosensors with Nanomaterials

Urea is a key indicator of renal function. Urease, an enzyme that hydrolyzes urea into ammonia and carbon dioxide, is commonly immobilized on nanomaterials to create sensitive biosensors [33].

- Nanomaterials like Ni-Co-MOFs, CNTs, and metal oxides provide ideal scaffolds for stabilizing urease [34].
- Immobilization improves enzyme activity retention and electron transfer efficiency, enabling accurate urea detection in biological samples.

7.4.2.2 Non-enzymatic Detection Methods Using Metal Oxides

Non-enzymatic urea sensing circumvents the instability and degradation issues associated with enzymes:

- Metal oxides (such as NiO, CuO, ZnO) exhibit intrinsic electrocatalytic activity toward urea oxidation [16, 35].
- These approaches are advantageous for achieving longer sensor lifetimes and operation under harsh conditions.

7.4.3 Creatinine Sensing

7.4.3.1 Functionalized Nanomaterials for Creatinine Detection

Creatinine is a crucial biomarker for assessing kidney function. Functionalized nanomaterials have improved creatinine detection through various strategies:

- Molecularly Imprinted Polymers (MIPs)**:** Artificial recognition sites are engineered for high-specificity binding to creatinine [36].
- Carbon-Based Nanomaterials: CNTs, graphene, and quantum dots provide high conductivity and surface functional groups to enhance sensor responses [37, 38].

7.4.3.2 Advances in Wearable and Point-of-Care (POC) Sensors

Recent developments focus on integrating nanomaterial-based sensors into wearable platforms and point-of-care devices:

- Wearable Patches and Contact Lenses: Flexible substrates embedded with nanomaterials allow real-time, noninvasive creatinine monitoring in sweat, tears, and interstitial fluid [39].
- Miniaturized POC Devices: Portable devices equipped with nanostructured electrodes offer rapid and accurate creatinine detection, facilitating early diagnosis and home monitoring [40].

7.5 Recent Advances and Applications

The field of nanomaterial-enabled biosensing for UA, urea, and creatinine has experienced significant growth, driven by the need for real-time, portable, and highly sensitive diagnostics. This section discusses the latest trends in wearable technologies, point-of-care systems, and the emerging roles of artificial intelligence (AI) and the Internet of Things (IoT) in nano-sensing [41].

7.5.1 Wearable and Implantable Sensors

7.5.1.1 Nanomaterials in Flexible and Miniaturized Biosensors for Real-Time Monitoring

Wearable and implantable sensors represent a frontier in noninvasive, continuous health monitoring. The integration of nanomaterials such as graphene, CNTs, MOFs, and conductive polymers has enabled the fabrication of flexible, lightweight, and highly sensitive devices.

- **Flexible Biosensors:** Materials like graphene and nanocomposite hydrogels are printed or embedded onto flexible substrates (e.g., textiles, soft patches, contact lenses) for detecting biomarkers from sweat, tears, or interstitial fluids.
- **Implantable Sensors:** Biocompatible nanomaterials allow for minimally invasive sensors that can operate inside the body, enabling long-term monitoring without causing significant immune responses.

For example: A Ni-Co-MOF hydrogel-based contact lens sensor was successfully developed for real-time urea monitoring in tear samples, demonstrating excellent flexibility, sensitivity, and biocompatibility [29].

7.5.2 Point-of-Care (POC) and Lab-on-a-Chip Devices

7.5.2.1 Integration of Nanomaterial-Based Sensors in Portable Devices

Point-of-care (POC) diagnostics aim to decentralize laboratory testing by enabling rapid analysis directly at the site of patient care. Lab-on-a-chip (LOC) devices, often microfluidic platforms, integrate multiple laboratory functions on a single chip.

- **Nanomaterial Enhancement:** Incorporating nanomaterials into LOC devices improves sensitivity, reduces sample volume requirements, and shortens response times.

- **Portability:** Miniaturized nanomaterial-based electrochemical and optical sensors have been developed for uric acid, urea, and creatinine detection using a few microliters of biological fluids.

For example: A portable smartphone based sensor using a CuO-ionic liquid-rGO composite allowed on-site creatinine testing with excellent accuracy, facilitating rapid diagnostics outside traditional labs [42].

7.5.3 Smartphone-Integrated and AI-Assisted Detection

The integration of smartphone platforms and AI algorithms with nanomaterial-based biosensors is revolutionizing personalized healthcare [43].

- **Smartphone Integration:** Optical and electrochemical signals from nanosensors can be transmitted to smartphones for instant analysis, data storage, and remote health management [42].
- **Artificial Intelligence (AI):** Machine learning algorithms analyze complex biosensor signals, improving the accuracy and predictive capabilities of biomarker detection.
- **Internet of Things (IoT):** Sensor networks connected through IoT allow continuous real-time health monitoring, cloud data storage, and healthcare provider alerts.

For example: Recent developments include smartphone apps coupled with nanomaterial-enabled biosensors for UA detection in saliva, using AI algorithms to distinguish between healthy and disease conditions based on real-time sensor data [44].

7.6 Challenges and Future Perspectives

Despite significant advancements in nanomaterial-based sensing technologies for uric acid, urea, and creatinine, several critical challenges must be addressed before these innovations can be fully translated into clinical and commercial applications [45].

A key concern is biocompatibility and potential toxicity; some nanomaterials, particularly metal nanoparticles such as AgNPs and quantum dots, may induce oxidative stress, inflammation, or cytotoxicity in biological systems. Strategies such as surface functionalization with biocompatible polymers (e.g., PEGylation) have shown promise in mitigating these effects, though thorough in vivo validation remains essential. Clinical-grade biosensors must also undergo stringent regulatory evaluation, following standards such as ISO 10993 to ensure safety and efficacy.

Current research efforts are therefore directed toward developing nontoxic, biodegradable, or "green-synthesized" nanomaterials for safer use in wearable and implantable biosensors [46, 47].

In addition to safety, scalability and commercialization present further barriers. While many nanomaterial-based sensors demonstrate excellent laboratory performance, mass production faces challenges including high fabrication costs, batch-to-batch variability, and difficulty in maintaining reproducibility. Integrating these materials into flexible or portable sensing platforms without compromising functionality remains technically complex. Furthermore, achieving regulatory approval from agencies such as the FDA or EMA requires extensive clinical validation and reliability testing, which can be both time-and resource-intensive. Efforts are underway to optimize scalable fabrication techniques, such as roll-to-roll printing, and to standardize nanomaterial characterization protocols to support regulatory compliance and industrial scaling [48, 49].

Looking ahead, future research is shifting toward next-generation materials and system-level innovations [50]. MXenes—two-dimensional transition metal carbides and nitrides—exhibit excellent electrical conductivity and hydrophilic surfaces, making them promising candidates for advanced biosensing. Similarly, MOFs provide high surface area and chemical tunability, suitable for both enzyme immobilization and non-enzymatic detection. Hybrid nanocomposites that combine materials like graphene-MOFs or carbon nanotube-metal oxide systems are also being explored to synergistically enhance sensitivity, selectivity, and operational stability.

The integration of artificial intelligence (AI) and machine learning with nanomaterial-based biosensors is emerging as a powerful approach for real-time data processing and smart diagnostics. Additionally, the development of self-healing, biodegradable sensors offer exciting potential for next-generation wearable and implantable platforms that are both sustainable and patient-friendly [51].

7.7 Conclusion

Over the past decade, the use of nanomaterials has transformed the landscape of biosensing for critical renal biomarkers like UA, urea, and creatinine. Through their remarkable properties—such as high surface area, superior electrical conductivity, tunable optical features, and catalytic activities—nanomaterials have significantly enhanced the sensitivity, selectivity, and stability of both electrochemical and optical biosensors.

Key advancements include:

- Electrochemical Sensors: Nanomaterial-modified electrodes demonstrated improved electron transfer rates and lower detection limits for rapid and precise analyte quantification.

- Optical Sensors: Fluorescence, colorimetric, and plasmonic techniques using quantum dots and noble metal nanoparticles allowed highly sensitive and visually intuitive detection mechanisms.
- Enzyme-Based and Non-enzymatic Sensors: Immobilization of enzymes on nanomaterials boosted biorecognition performance, while non-enzymatic sensors offered greater operational stability and longevity.
- Wearable, Point-of-Care, and Smartphone-Integrated Devices: Flexible, miniaturized, and portable systems employing nanomaterials have enabled real-time, user-friendly monitoring of biomarkers outside of traditional clinical settings.

These innovations have laid a strong foundation for more personalized, accessible, and real-time healthcare diagnostics.

7.8 Future Outlook

Looking ahead, nanotechnology is set to revolutionize biosensing in healthcare:

- Emerging nanomaterials like MXenes, MOFs, and hybrid composites will offer new functionalities, pushing sensor performance to unprecedented levels.
- Smart integration with AI, IoT, and machine learning platforms will enable biosensors to not only detect biomarkers but also predict health trends, personalize treatments, and alert healthcare providers automatically.
- Wearable and implantable biosensors powered by safe, biocompatible, and self-healing nanomaterials will make continuous, noninvasive health monitoring a reality.
- Commercial scalability and regulatory advances will bring affordable, reliable biosensing devices from the lab to bedside, home, and beyond.

Acknowledgments This book chapter has emanated from research supported under the European Union's Horizon 2020 research and innovation programme under the Marie Skłodowska-Curie grant agreement No 101034252.

References

1. Ma C, Jiang N, Sun X et al (2023) Progress in optical sensors-based uric acid detection. Biosens Bioelectron 237:115495. https://doi.org/10.1016/j.bios.2023.115495
2. Botewad SN, Gaikwad DK, Girhe NB et al (2023) Urea biosensors: a comprehensive review. Biotech App Biochem 70:485–501. https://doi.org/10.1002/bab.2168
3. Dong Y, Luo X, Liu Y et al (2022) A disposable printed amperometric biosensor for clinical evaluation of creatinine in renal function detection. Talanta 248:123592. https://doi.org/10.1016/j.talanta.2022.123592

4. Jayasekhar Babu P, Tirkey A, Mohan Rao TJ et al (2022) Conventional and nanotechnology based sensors for creatinine (a kidney biomarker) detection: a consolidated review. Anal Biochem 645:114622. https://doi.org/10.1016/j.ab.2022.114622
5. Li L, Wang T, Zhong Y et al (2024) A review of nanomaterials for biosensing applications. J Mater Chem B 12:1168–1193. https://doi.org/10.1039/D3TB02648E
6. Pechyen C, Tangnorawich B, Toommee S et al (2024) Green synthesis of metal nanoparticles, characterization, and biosensing applications. Sensors Int 5:100287. https://doi.org/10.1016/j.sintl.2024.100287
7. Hernández HD, Aparicio-Martinez E, Dominguez RB, Gutiérrez JM (2023) Development of non-enzymatic sensor for uric acid detection based on gold nanoparticles electrodeposited on laser-induced graphene electrodes. In: Trujillo-Romero CJ, Gonzalez-Landaeta R, Chapa-González C et al (eds) XLV Mexican conference on biomedical engineering. Springer International Publishing, Cham, pp 594–602
8. Shang K, Wang S, Chen S, Wang X (2022) Sensitivity detection of uric acid and creatinine in human urine based on Nanoporous gold. Biosensors 12:588. https://doi.org/10.3390/bios12080588
9. Barreto FC, Barberis ME, Mounienguet NK et al (2025) Biosynthesized silver nanoparticles anchored on a carbon material derived from maple leaves for the development of a green non-enzymatic biosensor for creatinine sensing. Green Anal Chem 13:100253. https://doi.org/10.1016/j.greeac.2025.100253
10. Ayyandurai N, Venkatesan S, Raman S (2024) Palladium nanoparticle-decorated copper-hemin metal organic framework for enzymatic electrochemical detection of creatinine in human urine. ACS Appl Bio Mater 7:8444–8455. https://doi.org/10.1021/acsabm.4c01285
11. Kumar P, Rajan R, Upadhyaya K et al (2024) Metal oxide nanomaterials based electrochemical and optical biosensors for biomedical applications: recent advances and future prospectives. Environ Res 247:118002. https://doi.org/10.1016/j.envres.2023.118002
12. George JM, Antony A, Mathew B (2018) Metal oxide nanoparticles in electrochemical sensing and biosensing: a review. Microchim Acta 185:358. https://doi.org/10.1007/s00604-018-2894-3
13. Pan Y, Zuo J, Hou Z et al (2020) Preparation of electrochemical sensor based on zinc oxide nanoparticles for simultaneous determination of AA, DA, and UA. Front Chem 8:592538. https://doi.org/10.3389/fchem.2020.592538
14. Fu L, Zheng Y, Wang A et al (2016) An electrochemical sensor based on reduced graphene oxide and ZnO Nanorods-modified glassy carbon electrode for uric acid detection. Arab J Sci Eng 41:135–141. https://doi.org/10.1007/s13369-015-1621-1
15. Bertel L, Miranda DA, García-Martín JM (2021) Nanostructured titanium dioxide surfaces for electrochemical biosensing. Sensors 21:6167. https://doi.org/10.3390/s21186167
16. Magar HS, Hassan RYA, Abbas MN (2023) Non-enzymatic disposable electrochemical sensors based on CuO/Co3O_4@MWCNTs nanocomposite modified screen-printed electrode for the direct determination of urea. Sci Rep 13:2034. https://doi.org/10.1038/s41598-023-28930-4
17. Khursheed S, Sarwar S, Hussain D et al (2025) Electrochemical detection of creatinine at picomolar scale with an extended linear dynamic range in human body fluids for diagnosis of kidney dysfunction. Anal Chim Acta 1353:343978. https://doi.org/10.1016/j.aca.2025.343978
18. Saddique Z, Faheem M, Habib A et al (2023) Electrochemical creatinine (bio)sensors for point-of-care diagnosis of renal malfunction and chronic kidney disorders. Diagnostics 13:1737. https://doi.org/10.3390/diagnostics13101737
19. Gao Q, Fu J, Xiong F et al (2024) A Multi-Channel urine sensing detection system based on creatinine, uric acid, and pH. Biosensors 14:473. https://doi.org/10.3390/bios14100473
20. Zalke JB, Bhaiyya ML, Jain PA et al (2024) A machine learning assisted non-enzymatic electrochemical biosensor to detect urea based on multi-walled carbon nanotube functionalized with copper oxide micro-flowers. Biosensors 14:504. https://doi.org/10.3390/bios14100504
21. Jeon SM, Lee HB, Ha CH et al (2024) High-performance electrochemical creatinine sensors based on β-Lead dioxide/single-walled carbon nanotube electrodes. Anal Chem 96:15941–15949. https://doi.org/10.1021/acs.analchem.4c02985

22. Neyra Recky JR, Montero-Jimenez M, Scotto J et al (2024) Urea biosensing through integration of urease to the PEDOT-polyamine conducting channels of organic electrochemical transistors: pH-change-based mechanism and urine sensing. Chem 12:124. https://doi.org/10.3390/chemosensors12070124
23. Williams RJ, Crapnell RD, Dempsey NC et al (2022) Nano-molecularly imprinted polymers for serum creatinine sensing using the heat transfer method. Talanta Open 5:100087. https://doi.org/10.1016/j.talo.2022.100087
24. Chen X, Wu Y, Qin Y et al (2025) Molecularly imprinted polyaniline-coated cu-Zeolitic Imidazolate framework nanoparticles: Uricase-mimicking "Polynanozyme" catalyzing uric acid oxidation. ACS Nano 19:9981–9993. https://doi.org/10.1021/acsnano.4c16272
25. Chakkarapani LD, Arumugam S, Brandl M (2021) Layer-by-layer sensor architecture of polymers and nanoparticles for electrochemical detection of uric acid in human urine samples. Mater Today Chem 22:100561. https://doi.org/10.1016/j.mtchem.2021.100561
26. Wen Y, Wang X, Zhao J et al (2025) Preparation and application of enzyme-based hydrogels. Biosensors and Bioelectronics: X 23:100594. https://doi.org/10.1016/j.biosx.2025.100594
27. Chakroborty S, Nath N, Mahal A et al (2024) Stimuli-responsive nanogels: a smart material for biomedical applications. J Mol Liq 403:124828. https://doi.org/10.1016/j.molliq.2024.124828
28. Thirumalai D, Santhamoorthy M, Kim S-C, Lim H-R (2024) Conductive polymer-based hydrogels for wearable electrochemical biosensors. Gels 10:459. https://doi.org/10.3390/gels10070459
29. Mukundan G, Badhulika S (2024) Nickel-cobalt metal-organic frameworks based flexible hydrogel as a wearable contact lens for electrochemical sensing of urea in tear samples. Microchim Acta 191:252. https://doi.org/10.1007/s00604-024-06339-8
30. Li F, He T, Wu S et al (2021) Visual and colorimetric detection of uric acid in human serum and urine using chitosan stabilized gold nanoparticles. Microchem J 164:105987. https://doi.org/10.1016/j.microc.2021.105987
31. Rezaei H, Jouyban A, Rahimpour E (2022) Development of a new method based on gold nanoparticles for determination of uric acid in urine samples. Spectrochim Acta A Mol Biomol Spectrosc 272:120995. https://doi.org/10.1016/j.saa.2022.120995
32. Singh G, Sharma S, Singh A et al (2024) Graphene oxide functionalized organosilanes for electrochemical detection of uric acid in urine and water. Mater Chem Phys 319:129347. https://doi.org/10.1016/j.matchemphys.2024.129347
33. Mashhadban KF, Gorgani L, Najafpour-Darzi G (2024) Enzymatic electrochemical biosensors for urea detection: a review. Sensors Actuators A Phys 374:115499. https://doi.org/10.1016/j.sna.2024.115499
34. Singh AK, Singh M, Verma N (2020) Electrochemical preparation of Fe3O4/MWCNT-polyaniline nanocomposite film for development of urea biosensor and its application in milk sample. Food Measure 14:163–175. https://doi.org/10.1007/s11694-019-00278-2
35. Amin S, Tahira A, Solangi A et al (2019) A practical non-enzymatic urea sensor based on $NiCo_2O_4$ nanoneedles. RSC Adv 9:14443–14451. https://doi.org/10.1039/c9ra00909d
36. Prabhu SN, Mukhopadhyay SC, Gooneratne CP et al (2020) Molecularly imprinted polymer-based detection of creatinine towards smart sensing. Med Devices Sens 3:e10133. https://doi.org/10.1002/mds3.10133
37. Bhatt P, Kukkar D, Yadav AK, Kim K-H (2024) Carbon dot-copper nanocomposite-based fluorescent sensor for detection of creatinine in urine samples of CKD patients. Spectrochim Acta A Mol Biomol Spectrosc 307:123666. https://doi.org/10.1016/j.saa.2023.123666
38. Ponnaiah SK, Prakash P (2020) Carbon dots doped tungstic anhydride on graphene oxide nanopanels: a new picomolar-range creatinine selective enzymeless electrochemical sensor. Mater Sci Eng C 113:111010. https://doi.org/10.1016/j.msec.2020.111010
39. Han S, Yamamoto S, Jung C-Y et al (2024) Wearable sensors for monitoring chronic kidney disease. Commun Mater 5:153. https://doi.org/10.1038/s43246-024-00606-0
40. Gonzalez-Gallardo CL, Arjona N, Álvarez-Contreras L, Guerra-Balcázar M (2022) Electrochemical creatinine detection for advanced point-of-care sensing devices: a review. RSC Adv 12:30785–30802. https://doi.org/10.1039/d2ra04479j

41. Wasilewski T, Kamysz W, Gębicki J (2024) AI-assisted detection of biomarkers by sensors and biosensors for early diagnosis and monitoring. Biosensors 14:356. https://doi.org/10.3390/bios14070356
42. Teekayupak K, Aumnate C, Lomae A et al (2023) Portable smartphone integrated 3D-printed electrochemical sensor for nonenzymatic determination of creatinine in human urine. Talanta 254:124131. https://doi.org/10.1016/j.talanta.2022.124131
43. Mishra A, Singh PK, Chauhan N et al (2024) Emergence of integrated biosensing-enabled digital healthcare devices. Sens Diagn 3:718–744. https://doi.org/10.1039/d4sd00017j
44. Liu W, Liu S, Fan K et al (2024) Machine-learning-based colorimetric sensor on smartphone for salivary uric acid detection. IEEE Sensors J 24:32991–33000. https://doi.org/10.1109/jsen.2024.3404646
45. Cánovas R, Cuartero M, Crespo GA (2019) Modern creatinine (bio)sensing: challenges of point-of-care platforms. Biosens Bioelectron 130:110–124. https://doi.org/10.1016/j.bios.2019.01.048
46. Hu C, Wang L, Liu S et al (2024) Recent development of implantable chemical sensors utilizing flexible and biodegradable materials for biomedical applications. ACS Nano 18:3969–3995. https://doi.org/10.1021/acsnano.3c11832
47. Alam F, Ashfaq Ahmed M, Jalal A et al (2024) Recent Progress and challenges of implantable biodegradable biosensors. Micromachines 15:475. https://doi.org/10.3390/mi15040475
48. Akhlaghi AA, Kaur H, Adhikari BR, Soleymani L (2024) Editors' choice—challenges and opportunities for developing electrochemical biosensors with commercialization potential in the point-of-care diagnostics market. ECS Sens Plus 3:011601. https://doi.org/10.1149/2754-2726/ad304a
49. Han X, Lin X, Sun Y et al (2024) Advancements in flexible electronics fabrication: film formation, patterning, and Interface optimization for cutting-edge healthcare monitoring devices. ACS Appl Mater Interfaces 16:54976. https://doi.org/10.1021/acsami.4c11976
50. Cho Y, Choi Y, Jang Y, Seong H (2025) Nanomaterial-enhanced biosensing: mechanisms and emerging applications. Adv Healthcare Mater. https://doi.org/10.1002/adhm.202500189
51. Sreejith S, Leo Joseph LMI, Kollem S et al (2023) Biodegradable sensors: a comprehensive review. Measurement 219:113261. https://doi.org/10.1016/j.measurement.2023.113261

Chapter 8
AI and IoT Integration in Biomolecular Sensing Systems

J. M. Radhika, J. Ajayan, Sandip Bhattacharya, and S. Sreejith

Abstract Biomolecular sensing devices have gained significant appeal as a means of gathering proof of concept data due to their exceptional selectivity and sensitivity. Point-of-care (POC) biosensing combined with artificial intelligence (AI) and IoT (Internet of Things) has the potential to completely transform diagnostic techniques by providing quick, precise, and easily accessible health evaluation at the patient level. By delivering advanced diagnostic abilities straight to the patient at home, in the field, or by their bedside, POC bioanalysis is continuing to transform patient care. Numerous sensing techniques make these systems possible, providing sensitive and precise information quickly without the need for testing that is centralized. This chapter thoroughly and critically examines the role and significance of AI and IoT in enhancing the functionality, detection precision, decision-making capacity, and assessment of various bio molecular sensing devices. Recent advances of AI in enhancing the functionality of Internet of Medical Things (IoMT) and POC devices utilized in state-of-the-art medical domains like diabetes treatment, cancer detection, and cardiac assessment is also covered in this review.

Keywords Point of care devices · Biomolecular sensors · AI · IoT · IoMT · Explainable AI

8.1 Introduction

The Internet of Things (IoT) technological subset referred as IoMT is made up of medical equipment that are network-connected and used for monitoring healthcare. In order to give healthcare monitoring without requiring human intervention, IoMT

J. M. Radhika (✉) · J. Ajayan · S. Bhattacharya
SR University, Warangal, Telangana, India

S. Sreejith
New Horizon College of Engineering, Bengaluru, Karnataka, India

R. Naik et al. (eds.), *Biocompatible Nanomaterials for Smart Sensors*, Nanostructure Science and Technology,
https://doi.org/10.1007/978-3-032-01745-1_8

devices, also referred to as healthcare IoT combine interfacial sensors, machine learning (ML) which is AI based and automation. IoMT technologies enable health parameters wireless detection, which helps to decrease needless hospital visits and the related health expenses. IoMT technology allows remote access to collect, process, and transmit medical data via a secure network by connecting clinicians and patients through medical devices. AI is a modern approach based on computer science that develops programs and algorithms to give machines intelligence and efficiency so they can perform tasks that frequently need highly skilled human intelligence [1–10]. A variety of subsets of AI such as conventional neural networks, deep learning (DL), ML, fuzzy logic, and speech recognition, each have unique powers and features that can improve the productivity of modern medical sciences. Such intelligent systems simplified human intervention in decision-making, medical imaging, and clinical diagnostics. IoMT technology provides bio-analytical cutting-edge tools that combines software and network-connected medical equipment to improve human health. Digital healthcare is putting automatic detection tools and integrated cloud goods in consumers' hands, while conventional medical surveillance is undergoing a radical transformation. This technological revolution enables patients, doctors, and rural communities to access high-quality healthcare services in order to improve outcomes. Users may monitor their health with POC equipment like thermometers, ultrasonography, ECG readers and glucometers, and because they possess Internet connectivity and the ability to save data in the cloud. The term infectious disease describes the functional impairment of health brought on by a variety of pathogenic microbes such as viruses, bacteria, parasites, and fungi that are present inside the body. Rising rates of infectious diseases brought on by pathogens are an indication of the current pandemic on a worldwide scale. In a relatively short period of time, these diseases are spreading more quickly, endangering the general well-being of the populace and maybe the economy. A fundamental obstacle to medical care is the insufficiency of traditional diagnostic instruments in comparison to the laborious and intricate laboratory-based diagnosis procedure. In order to diagnose infectious diseases quickly and provide on-site outcomes that support timely and early intervention for better care, point-of-care testing, has been developed in the modern period. Furthermore, POC testing devices are essential for stopping the infectious illnesses spread since they provide real-time testing and lab-quality microbiological identification in a few minutes. Infectious disease outbreaks can be contained with the help of improved treatment and prompt diagnosis. Currently, technical advancements in the IoMT field are being used to facilitate such point-of-care testing [11–20]. The IoMT allows point-of-care testing devices to operate wirelessly and link to medical centers and health experts. Because they interact with and detect target analytes using biorecognition elements, biosensors are bioanalytical systems. The molecule of interest or analyte, that the biosensor detects; the structure or recognition element or molecule or that interacts with the analyte; the procedure or transducer or apparatus that converts the interaction between the recognition element and analyte into a readable signal; and the data handling procedures, which allow for the analysis and interpretation of the generated data, are the four main parts of a typical biosensor (Fig. 8.1).

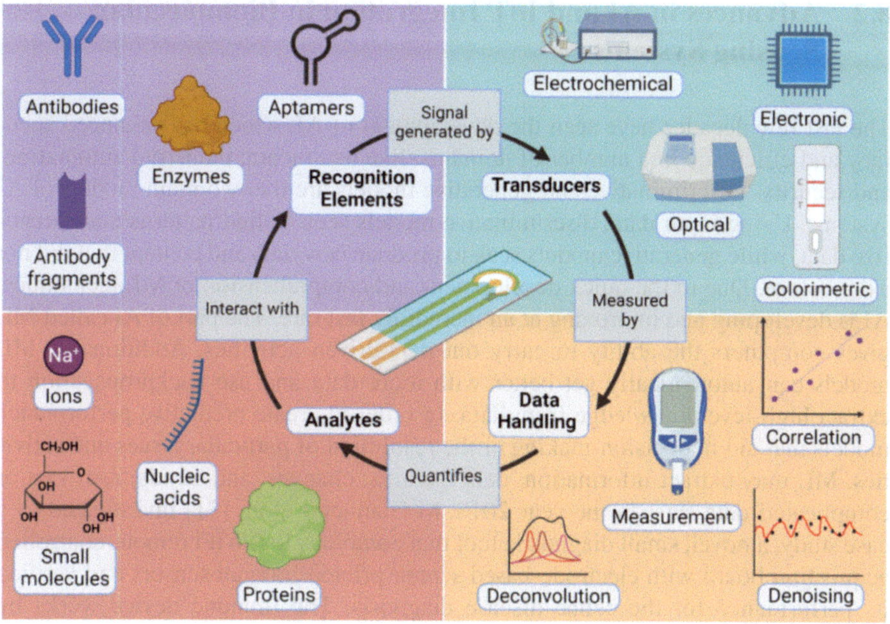

Fig. 8.1 The process of developing biosensors (Reprinted with permission from MDPI) [21]

By providing quick, precise, and easily accessible health evaluation at the patient level, the incorporation of AI and IoT into POC biosensing holds the potential to completely transform diagnostic techniques. POC sensing technologies have significantly improved medical diagnostics' accessibility and speed, but they frequently fall short of the precision and dependability of conventional lab-based testing. This flaw is mostly caused by intrinsic restrictions of POC devices' functioning and design. These gadgets usually focus on single-target detection and employ simplified techniques with fewer sample numbers, which could lead to less precise results. Additionally, compared to the more advanced equipment utilized in laboratories, the transducers in POC devices typically provide signals with reduced sensitivity, in part due to the volume of background noise. Further variability that could jeopardize test results is introduced when POC equipment are used in uncontrolled settings, including homes or fields, where humidity, temperature and operator handling abilities can vary. But as AI continues to progress, POC technologies appear to have a bright future. Through speeding up the creation of biomarkers and receptors, improving data processing skills, increasing device accuracy, and streamlining the operational difficulties of POC tests, AI has the ability to address these issues. POC diagnostics' performance and dependability can be significantly enhanced by combining AI with biomolecular sensing technologies, bringing them closer to laboratory standards. This review focuses on emerging IoMT and AI based advanced technologies used in various bio-sensing platforms [22–30].

8.2 Advances in AI and IoT Integration in Biomolecular Sensing Systems

The last few decades have seen the development of AI, which has enhanced accuracy and efficiency in a number of domains, like healthcare, industrial automation, and security. Discriminative and generative models are the two main models of AI systems. Using input data, discriminative models seek to find patterns and categorize data, while generative models seek to produce new data and content that did not exist before. Due to the quick development and comprehension of ML techniques, AI is developing and improving at an incredibly fast rate. The part of AI called ML gives computers the ability to carry out data-driven activities. Additionally, ML models can automatically get better with more data and use backpropagation to extract high-level knowledge from data. In order to create predictive performance models that aid in decision-making or the resolution of particular issues and activities, ML may extract information, data features, changes, and trends from vast or complicated data sets. In the year 2018, M.Giannetto et al. [31] investigated as a case study, a novel, small diagnostic tool that combines IoT-WiFi remote-controlled acquisition board with electrode-based screen-printed immunosensors and verified its performance for the celiac disease diagnosis. The immune device works by allowing the open tissue enzyme transglutaminase to chemisorb onto the surface of carbon screen-printed electrodes functionalized with nanoparticles of gold. The immobilized bioreceptor identifies anti-tissue IgA and IgG trans-glutaminase target antibodies as highly specific indicators associated with celiac disease. A specially designed WiFi IoT integrated board receives and processes the amperometric sensor signal, enabling sharing of data in real time on cloud services to immediately alert all users like caretakers, physicians, etc., on device outcome (Fig. 8.2). Neither specialized software nor hardware was needed for the reported solution. To get limit of quantitation (LOQ) and limit of detection (LOD) levels that are diagnostically useful (IgA LOD = 3.2 AU mL^{-1}; IgG LOD = 1.4 AU mL^{-1}; IgA LOQ = 4.6 AU mL^{-1}; IgG LOQ = 2.3 AU mL^{-1}) and excellent midrange precision (less than 5% RSD), the immunosensors' analytical capabilities were enhanced by means of experimental design. The IoT-Wi-Fi device was suitable for diagnostic purposes due to its high discrimination capabilities between negative and positive serum control, which showed extraordinary statistical significance ($p < 0{,}001$).

In the year 2024, H. Q. Nguyen et al. [32] reported rapid CoV-2-SARS detection with a new POC platform that combines IoT technology with a 3D all-in-one printed microfluidic cartridge. Using an IoT-based diagnostic tool and a 3D-printed self-contained microfluidic cartridge, a novel molecular diagnostic POC platform was developed for CoV-2-SARS detection. The platform facilitates automated downstream procedures such as RNA extraction using magnetic beads, precise aliquoting of liquid into 4 micro-reactors, temperature regulation for an isothermal amplification reaction mediated by a reverse transcription loop, fluorescence real-time detection using a CMOS sensor, as well as data processing and visualization. With a detection limit of 1×10^{-1} copies/μl and a small size (height: 16 cm; width: 12 cm;

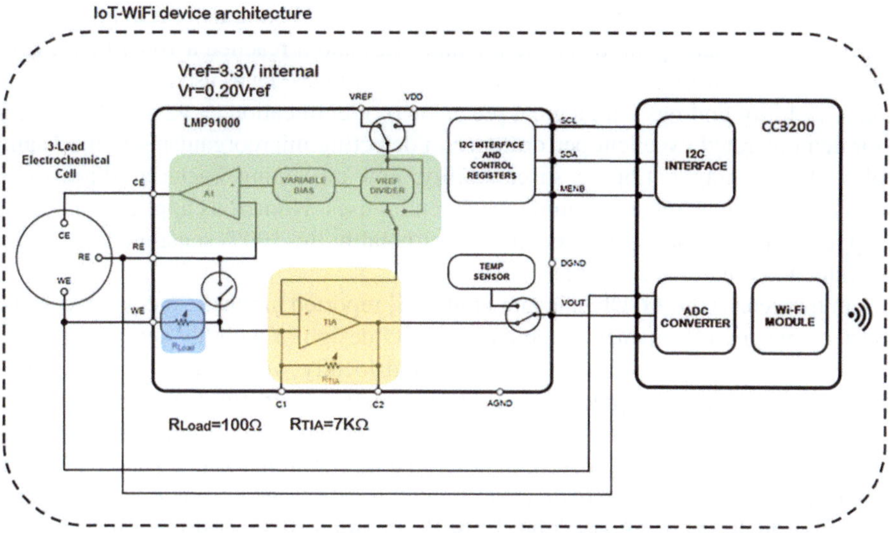

Fig. 8.2 IoT WiFi device architecture [31] (Reprinted with permission from Elsevier)

length: 9 cm), the entire procedure was finished in 70 min. This cutting-edge POC platform can be extremely helpful in resolving shortages in medical supplies and for COVID-19 testing because it is easy and inexpensive to manufacture using 3D printing technology. In the year 2024, H. Kobayashi et al. [33] reported utilizing transcriptome promoter-level analysis and AI to find a biomarker that may be a predictive predictor of cardiac problems in male Fabry disease patients. The most common lysosomal illness in Japan is called Fabry disease, and it is characterized by a broad range of symptoms. Adult-onset Fabry disease is mostly connected with three main concerns that need to be avoided: cardiac, renal, and central nervous system issues. Severe arrhythmia, cardiac muscle fibrosis, and cardiomyopathy are among the most frequent reasons of death for patients with Fabry disease. H. Kobayashi et al. used RNA extracted from patient venous blood for gene expression cap analysis (CapeAGE) to forecast Fabry disease's cardiac consequences. Likelihood ratio tests were also conducted for every dataset of RNA expression obtained from people without and with heart problems and examined the relationship between RNA expression and heart functional characteristics as determined by AI algorithms extracted from magnetic resonance imaging data. It was observed that male Fabry disease patients with heart problems had significantly greater levels of CHN1 expression, which may be linked to a variety of cardiac functional variables. Chimererin 1, a protein that activates GTPase and is unique to the GTP-binding Rac protein (involved in the development of cardiac fibrosis and oxidative stress), was encoded by CHN1. In the year 2024, R. Cui et al. [34] reported colorimetric smartphone-based biosensor with AI support for the visual, quick, and sensitive identification of harmful microorganisms. The reported study developed a colorimetric smartphone-based biosensor with AI support for the POC pathogenic bacteria detection that produce HAase. It was able to distinguish between different bacterial kinds and quantities in serum and PBS, and it reached a 10 CFU/mL LoD in 60 min. A YOLOv5 self-developed algorithm (Fig. 8.3) was used to analyze color data, yielding findings with a great precision of identification (> 92%). The biosensor demonstrated excellent performance in detecting microorganisms in food samples, determining antibiotic susceptibility, and evaluating bacterial illnesses in clinical settings, indicating numerous potential uses. Numerous applications of the suggested biosensor were successful, most notably the 100% sensitive assessment of infections in clinical samples.

In the year 2024, G. Dhandapani et al. [35] reported graphene metasurfaces and surface plasmon resonance in the analysis and design of a highly sensitive protein identification using a terahertz biosensor and locally weighted AI-assisted linear regression for behavior prediction. Protein detection is crucial for a variety of biological and biochemical fields, such as environmental monitoring, pharmaceutical research, and disease diagnostics. In order to uncover biomarkers linked to different medical disorders and to explain intricate biological processes, precise and effective detection techniques are essential. In order to attain high sensitivity, the reported study discusses the development and evaluation of an advanced terahertz protein detection biosensor that combines graphene metasurfaces and surface plasmon resonance. The sensor concept uses a graphene sheet as the active sensing element and uses a silica substrate to combine circular and rectangular ring resonators. Finite element simulations are used to optimize the sensor's material and geometric characteristics. With an ideal 508 GHzRIU^{-1} sensitivity and 5.089 quality factor, the

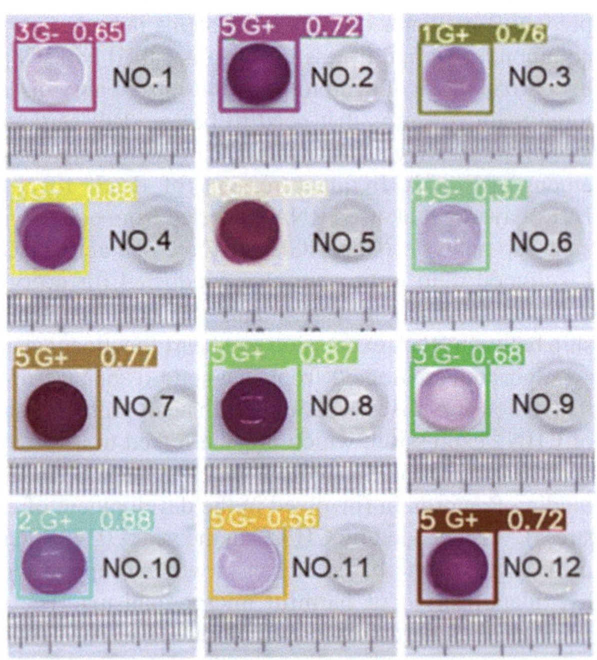

Fig. 8.3 The YOLOv5 algorithm's output and the 12 samples that the biosensor detected [34] (Reprinted with permission from Elsevier)

improved sensor design exhibits remarkable performance features. Further analysis revealed 11.293 RIU^{-1} high figure of merit, 0.607 low detection limit, and 0.022 decent signal-to-noise ratio. Additionally, the sensor showed potential for two-bit encoding applications. R2 values of up to 100% are obtained by analyzing and forecasting the sensor's performance over a range of parameter combinations using a locally weighted linear regression model. The suggested biosensor has potential for a number of uses, such as improving environmental monitoring, guaranteeing food safety, and improving medical diagnostics.

In the year 2025, M. Papa et al. [36] reported AI-enabled hyperspectral microscopy to identify *E. coli* that are alive but not culturable due to low-concentration antimicrobials. Under sublethal stress, such as antibiotic resistance, bacteria can evade detection by entering a viable but nonculturable (VBNonC) state; yet, timely identification of bacterial infections is essential for public health and food safety. These bacteria are still alive but cannot be found using conventional culture-based techniques without significant enrichment, hence sophisticated detection techniques are required. M.Papa et al. created a hyperspectral microscope imaging (HMicroI) framework with AI capabilities for quick VBNonC identification when using low-level antibiotics. The goals were exposing Escherichia coli K-12 to certain antibiotic stresses to produce the VBNonC state, obtaining HMicroI data documenting physiological alterations in VBNonC cells, and automating normal and VBNonC cell classification using deep learning image classification. Plate counting and live-dead staining indicated that the VBNonC condition was caused by oxidative low-level (hydrogen peroxide—0.01%) and acidic (peracetic acid—0.001%) stresses for 3 days (Fig. 8.4). Three distinctive spectral wavelengths were used to extract the spectral and spatial data from HMicroI into pseudo-RGB images. These RGB pseudo images were used to train an EfficientNetV2-based convolutional neural network architecture, which outperformed the model built on RGB images at 83.3% with a 97.1% VBNonC classification accuracy ($n = 200$). The findings demonstrate the possibility of quick, automated VBNonC detection with AI-enabled

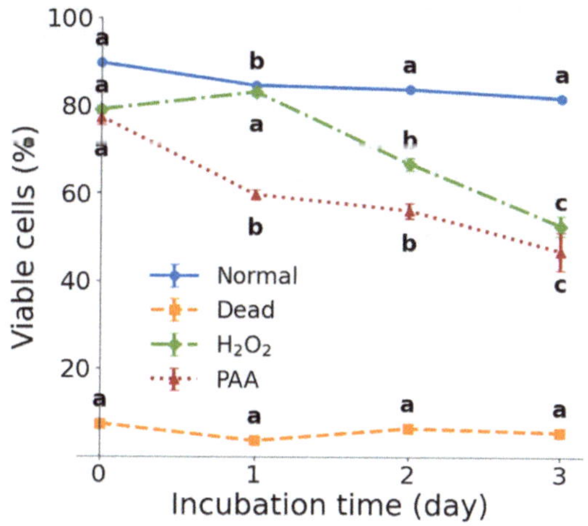

Fig. 8.4 Percentage of viable cells vs. incubation time as determined by live-dead staining [36] (Reprinted with permission from Elsevier)

hyperspectral microscopy, which could help with prompt action to stop foodborne outbreaks and diseases. In the year 2025, J.Y.Choi et al. [37] reported explainable AI (EXAI) powered prostate cancer detection with a dual-gate FET biosensor based on exosomal multi-markers. Because of its remarkable diagnostic efficacy, the Prostate Imaging Reporting and Data System (PI-RADataS) score, a mechanism for reporting cases of prostate MRI, has evolved into a routine prostate cancer (ProCa) screening technique. However, due to PI-RADataS's high false-positive rate and diagnosis accuracy of just 30–40% at most, PI-RADS 3 lesions represent an unmet medical need. ProCa screening system based on EXAI that incorporates a multi-marker biosensor based on a dual-gate sensitive field-effect transistor (DGFET) for the diagnosis of ambiguous lesions was reported (Fig. 8.5). By examining three urinary exosomal biomarkers' sensing patterns, this approach generates results that are easy to understand and gives clinicians the chance to make an evidence-based forecast. With a noninvasive approach and 102 blinded samples, findings demonstrated a high accuracy of the EXAI-based ProCa screening system with an AUC of 0.93. The accuracy of ProCa diagnosis in patients with PI-RADataS three was remarkably more than double that of patients with traditional PI-RADataS scoring. Additionally, this system offered a plausible justification for its conclusion that the TMEM256 biomarker is the primary determinant for PI-RADataS three screening. Research suggested that EXAI can support well-informed choices that are informed by knowledge of the importance of clinical considerations and visualized multi-biomarkers. Healthcare practitioners can receive practical, evidence-based ProCa diagnoses with the help of the EXAI-based sensor system.

In the year 2025, Yu Yang [38] investigated the use of wearable technology with AI sensors for tracking human health in sports. AI-based wearable sensor devices have a lot of promise and a wide range of development opportunities for sports and human health monitoring. Using sensor technologies such as accelerometers, wearable technology can measure the frequency and amplitude of acceleration during human movement and accurately record acceleration variation in all directions. Human movements' efficacy and quality can be assessed and examined using these data. By integrating with Web servers and database servers, sensor wearables can transmit data to the server for processing and archiving. AI technologies can then

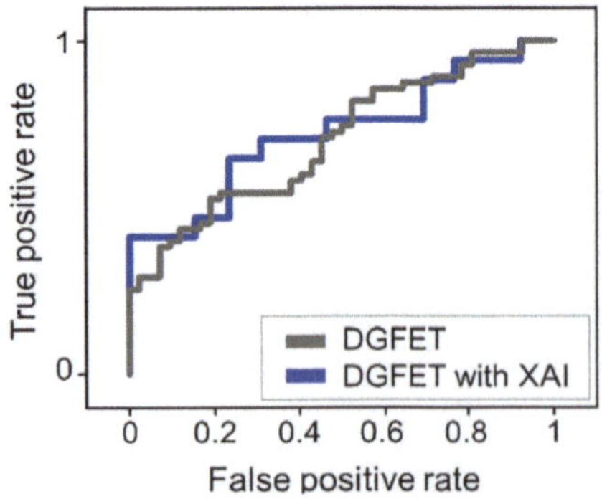

Fig. 8.5 ROC curves illustrating how well PI-RADataS results match dual-gate FET sensing data [37] (Reprinted with permission from Elsevier)

analyze and show data online in real time. These features offer individualized health monitoring services and can assist users in understanding their own athletic and medical status. As AI and sensor technologies continue to advance, sensor wearables will become more prevalent in the area of sports health surveillance. Future sensor wearables will be able to do more than just track fundamental metrics like acceleration; they will also be able to integrate other features like heart rate tracking and sleep analysis to offer more complete health monitoring services. By combining cloud computing, smartphones, and other technology, sensor wearable devices can also offer consumers auxiliary training functions and greater sports data analysis, assisting users in managing and improving their health. The deployment of sensor wearable devices still faces certain technical difficulties and privacy security concerns, but these will be resolved with the ongoing development of technology and sensible specification. In the year 2025, J.H.Wang et al. [39] reported verification of a predictive biomarker based on AI in patients with oligometastatic prostate cancer that is sensitive to castration. A multicenter retrospective research was done to assess the predictive and prognostic efficacy of a multimodal AI biomarker (MultiMAI; the ArteraAIProstateTest) among male omCSPC patients (n = 222). There were also 51 patients from the STOMP and ORIOLE phase two clinical trials in the group, which assigned patients to observation as opposed to MD-Therapy (metastasis-directed therapy). Digitalized histopathology slides and clinical variables were used to calculate MultiMAI scores. The complete cohort's time to castration-resistant and overall survival to prostate cancer were evaluated starting at the diagnostic time. Time of randomization was used to measure the trial cohort's metastasis-free survival. Patients having a high MultiMAI score throughout the entire cohort had shorter TTCRPC and much worse overall survival. The only variable that was substantially correlated with OS in a multivariate Cox model was the MultiMAI score. Patients with a high MultiMAI score in the group of patients randomly assigned to the ORIOL and STOMP studies had better metastasis-free survival with MD-Therapy ($p = 0.039$) than patients with a low score ($p = 0.04$). In the year 2025, Rebeka Torok et al. [40] reported analyzing serum N-glycome with the aid of AI to forecast how well chemotherapy will work for lung cancer. This reported new method for forecasting the effectiveness of chemotherapy for lung combines data analysis based on AI with patient serum N-glycome monitoring. Thirty-three patients with lung cancer receiving chemotherapy participated in the trial. Before and after the treatment, serum samples were collected. Following their enzymatic release, oligosaccharides N-linked were fluorophore-labelled and subjected to laser-induced fluorescence detection for analysis using capillary electrophoresis. The final model, which was refined to increase the classification accuracy, was built using QDA (Fig. 8.6). AI-based classifiers, which divide the data into binary classifications using a ML algorithm, carefully processed and assessed the resultant electrophorograms. The results of the chemotherapy treatments were strongly correlated with the structural alterations in the N-glycans, according to the classifier analysis approach (ROC >0.9).

This innovative fusion of AI and bioanalytical techniques produced a quick and accurate tool for estimating chemotherapy efficacy. This collaborative method not

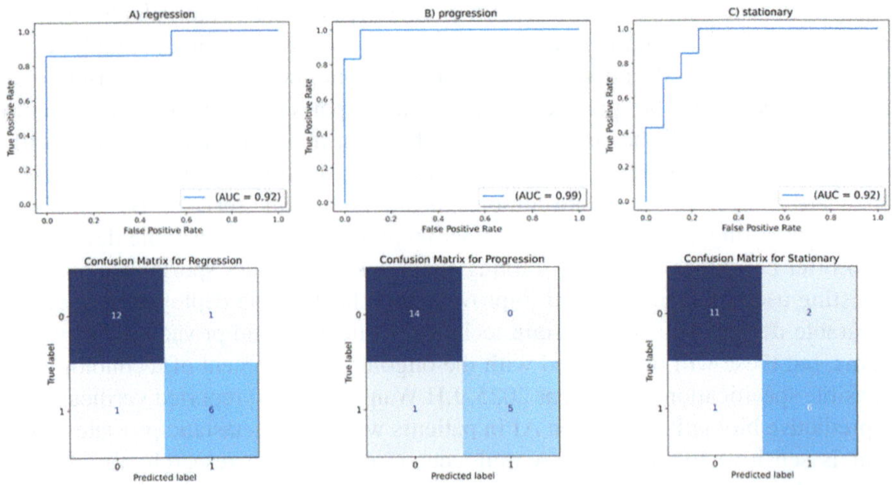

Fig. 8.6 Confusion matrix and ROC of QDA fine-tuned algorithm [40] (Reprinted with permission from Elsevier)

only showed encouraging outcomes in predicting the effectiveness of chemotherapy using N-glycan analysis, but it also proved the revolutionary potential of AI in biological sample analysis. The study may be limited by the patient population's heterogeneity with regard to histological type, stage of lung cancer, and kind of chemotherapy treatment. The comparatively modest sample cohort size, the incorporation of more clinical data, and enhancing model interpretability present further difficulties.

In the year 2025, S. Park et al. [41] reported using the multi-algorithm sequential reclassification technique for precision screening to facilitate needless bladder cystectomies. When detected at an advanced stage, bladder cancer frequently requires invasive intervention. This has led to the active use of AI-based precision screening and noninvasive cancer detection with biosensors. False negatives, the incorrect categorization of cancer patients as healthy, remain a serious worry, nonetheless, as they may have catastrophic consequences for their longevity. Additionally, whereas group methods such as soft voting and others can decrease misclassification and increase model accuracy, their usefulness is restricted and not suitable for all diagnostic applications. Utilizing an electrical sensitive urine biosensor, a cancer double-stage screening system was developed using an EXAI interpretation and AI model. Recognized for its substantial recurrence and high likelihood of developing from muscle noninvasive tumors to muscle-invasive tumors, the reported method was intended for the screening of bladder cancer. Patients received primary screening using the classification CatBoost model after 4 urinary biomarkers (PD-L1, CK18, CK8 and PD-1) were detected by a field-effect transistor biosensor. Additionally, patients' age and sex were recorded. Applying neural network-based local explanations, which provide an alternative viewpoint to CatBoost, patients who were first labelled as normal were reclassified. After a second round of screening, all of the false negatives from the primary screening could be correctly identified as cancer symptoms. Additionally, the development of the AI model is guided by global explanation in order to get high accuracy when trained on a suitable set of biomarker features (Fig. 8.7). In the year 2025, J. S. Y. Low et al. [42] reported an AI-assisted paper-based microfluidic multiplexed surface-enhanced Raman scattering (SERS) biosensor with electrophoretic removal and electrical modulation for precise acute myocardial infarction (AMyoI) prognosis and detection. Single molecules can be detected with remarkable sensitivity using SERS. In the multiplexed detection of HDL, LDL, and cTnI in human serum at a low LoD, an externally applied electric field that enables electrical modulation and electro-migrates unbound SERS tags without requiring multiple washing steps has been successfully developed and demonstrated to increase the biosensor's sensitivity and selectivity. This approach addresses the selectivity problem that point-of-care faces. Nonspecific binding's impact can be enhanced by nonspecifically absorbed species, which can obscure overlapping analyte peaks and be detected by ultra-sensitive detectors. Antifouling molecules have the ability to stop random adsorption on the surface of the sensor, although this does not totally eradicate it. The results demonstrate that nonspecifically bound species can be electromigrated by an electrically controlled device without interacting with indigenous albumin proteins. With an RSD of 10%, stability, repeatability, and reproducibility were all observed to be good. SERS spectra of high-dimensional fingerprints were interpreted and analyzed by AI utilizing feature selection and dimensionality reduction to accurately diagnose and prognosticate acute myocardial infarction. These ML techniques enable low RMSE quantification of LDL, cTnI and HDL biomarkers. Strong AUROC values of 0.950 ± 0.111 and 0.884 ± 0.139 were demonstrated using ML classifiers for the detection of early and recurrent AMyoI, respectively. A high 99% negative prediction value or more signifies a successful early rule-out of AMyoI. This study demonstrated that a straightforward, inexpensive, electrophoretic regulated biosensor using machine learning can identify, rule out, and anticipate recurrent AMyoI.

In the year 2025, F, Dai et al. [43] reported a small, isothermal, palm-sized fluorescence diagnostic intelligent IoT device that uses a 1-step/1-tube CRISPR- LAMP assay for personal health monitoring. At-home nucleic acid testing kits that are sensitive, accurate, and easy to use are becoming more and more necessary due to the occasional emergence of various infectious diseases and the growing demand for a better quality of life. WeD-mini (Fig. 8.8), a small, 61 g, palm-sized, isothermal

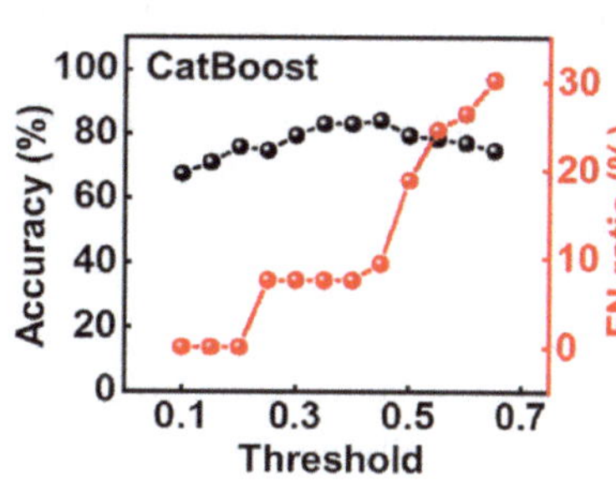

Fig. 8.7 Accuracy and false negative vs. threshold plot [41] (Reprinted with permission from Elsevier)

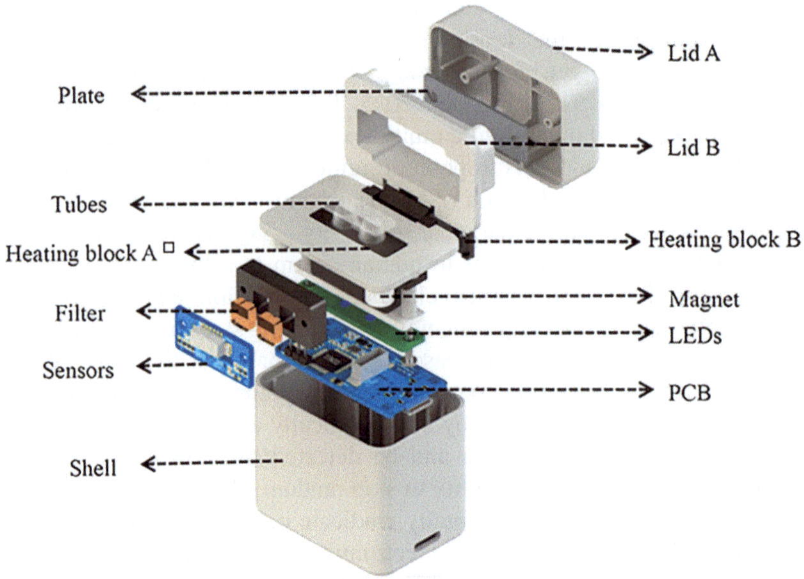

Fig. 8.8 WeD-mini, a small, 61 g, palm-sized, isothermal fluorescence diagnostic IoT gadget [43] (Reprinted with permission from Elsevier)

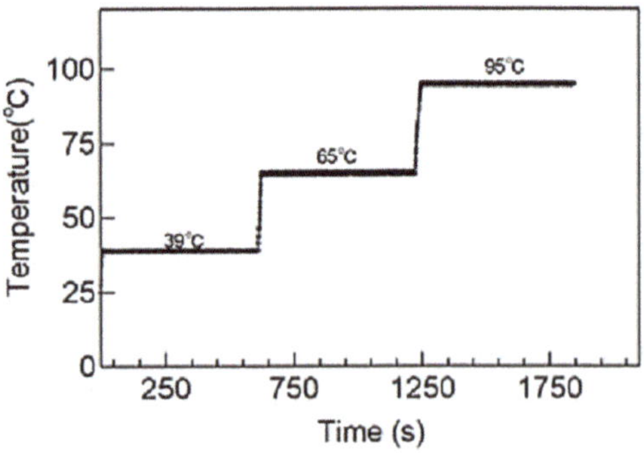

Fig. 8.9 WeD-mini temperature stability plot [43] (Reprinted with permission from Elsevier)

fluorescence diagnostic IoT gadget was introduced. The WeD-mini features an intelligent working algorithm, a simple industrial design, ultra-low power consumption, and a recently constructed, very sensitive optical sensor technology. The results can be readily uploaded to the "EzDx Cloud" for spatiotemporal illness mapping and general health management. Real-time fluorescence detection and smartphone-based automatic result interpretation were combined. The device supports a range of assays that operate at different temperatures (Fig. 8.9) and with varying fluorescence emission intensities, such as RPA (low intensity, 39 °C), LAMP (high intensity, 65 °C), and LAMP-PfAgo (high intensity, 65/95 °C), all while maintaining precise temperature control and exceptional fluorescence detection sensitivity.

Furthermore, to enhance LAMP compatibility and construct a dependable 1-step/1-tube CRISPR-LAMP assay, a more thermostable AapCRISPR-Cas12b variant was developed, which operates effectively at 63 °C. With the developed 1-step/1-tube CRISPR-LAMP assay, the WeD-mini demonstrated 100% sensitivity and specificity, making it suitable for at-home testing of influenza and SARS-CoV-2 viruses. Additionally, the WeD-mini exhibits potential uses in wastewater surveillance, pet respiratory disease monitoring, and meat adulteration detection, making it appropriate for a variety of private and public health use cases. In the year 2025, Y.Wang et al. [44] reported an AI-enabled bioinspired microbial taste chip with extremely quick reaction times and high selectivity. Since global water pollution is becoming a serious problem harming human health, real-time monitoring of water pollution is essential. Because of their quick reaction time and real-time monitoring capabilities, microbial taste chips show promise for monitoring water contamination. Even with the use of gene-editing techniques, sensor selectivity, the most important parameter, remains an unresolved issue despite the fact that there have been many research works on microbial taste chips. The response time was quite lengthy, taking at least three minutes. Bioinspired wireless microfluidic microbial taste device exhibited great selectivity and was enabled by AI that solves the selectivity constraint. It was observed that by using gated recurrent unit-based DL algorithms to harness the different patterns of the devices' temporal output current to various contaminants, 98.9% classification accuracy for Pb^{2+}, Cr^{6+} and Cu^{2+}, 3.75 times faster than the fastest previously known counterpart, the shortest response time of 48 s can be attained. High accuracy and linearity real-time sensing of Cu^{2+}, Pb^{2+}, and Cr^{6+} was made possible by the device. The chip's small size and wireless connectivity could help with global efforts to combat water pollution and find uses in quantitative real-time heavy metal ion monitoring in water.

In the year 2025, S. Harnsoongnoen et al. [45] reported fingertip glucose solution concentration and blood sugar level detection and classification via deep learning methods and a unique planar microwave sensor. Having undergone extensive design, fabrication, and testing to measure critical parameters like S_{11} (Fig. 8.10) and F_r, the Ω-shaped resonator (OSR) - split-ring resonator (SRR) microwave sensor has proven its capacity to accurately detect and distinguish glucose concentrations ranging from 0 to 200 mg/dL. The sensor was replicated using dry film and printed circuit boards processes to assess how relative permittivity and the thicknesses of the skin and blood layers affected its response, demonstrating its versatility in a range of scenarios. By combining a transmission line, OSR, and SRR, its innovative design maximizes electromagnetic field uniformity and concentration, aspects not seen in earlier designs. The sensor can detect and classify 3–5 different glucose levels within a single structure, and it can measure glucose solutions with outstanding sensitivity, especially in microfluidic channels. Furthermore, by addressing the impact of the fingertip's electrical and physical characteristics on its functionality, a crucial element not previously investigated, the sensor demonstrates its adaptability. The sensor solves issues with nonlinear response behavior by utilizing cutting-edge ML approaches, such as K-means unsupervised clustering and artificial neural networks, guaranteeing precise and trustworthy categorization of glucose. These

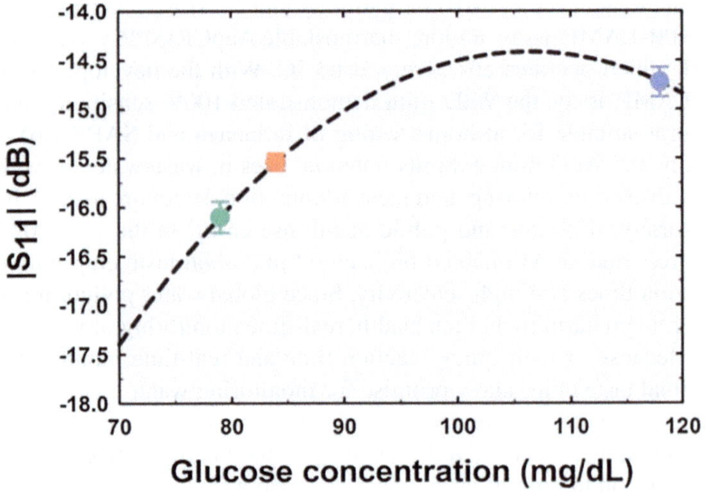

Fig. 8.10 Variation in S_{11} with glucose concentration [45]. (Reprinted with permission from Elsevier)

advancements, in addition to the sensor's compact size, reasonably priced design, and enhanced sensitivity over earlier microwave sensors, show how promising it is as a real-time, noninvasive glucose monitoring tool and for broader applications in chemical analysis and medical diagnostics.

8.3 Future Prospects

Biosensing has recently taken on a new dimension as a result of the promotion and application of concepts such as AI, big health, IoT, and big data. The development of wearable biosensors aims to overcome the limits of centralized, reactive health-care by providing individuals with knowledge into their own physical dynamics. However, complicated causality makes output analysis extremely challenging. The goal of tailoring the best precision medicine treatment for every patient depends on wearable sensors and AI. Better patient data collection and wearable sensor design to monitor wearers' environment, health, and fitness are made possible by combining these two fields. As the Internet of Things, big data, and big health move from concept to reality, AI-biosensors with the proper technical attributes face new opportunities and problems. It is feasible to bridge the gap between data collection and processing and improve the accuracy of diagnosis and therapy by integrating biosensors with AI methods like pattern recognition and classification algorithms. With their combination of sophisticated machine learning algorithms and wireless biosensing technologies, AI-biosensors show enormous potential for enabling cloud-connected POC diagnostics and ongoing healthcare monitoring [46–60].

8.4 Conclusion

Biosensing has gained new significance due to recent developments in the application and promotion of big data, IoT, and big health ideas. With its cutting-edge approaches to clinical support, predictive modeling, and diagnostics, AI has completely transformed the healthcare industry. By enabling the development of sophisticated medical detection and healthcare analytics, AI algorithms and data science have revolutionized the delivery of healthcare services and improved patient outcomes. Also development in wearable biosensor aims to overcome the drawbacks of reactive, centralized healthcare by empowering individuals with knowledge about their own body dynamics. However, because causation is complex, output analysis is quite challenging. The goal of customizing the optimal precision medicine treatment for each patient depends on wearable sensors and AI. In this chapter we have studied a few state-of-the-art advancements in AI and IoT biosensing systems.

References

1. Teodoro KBR, Pereira TS, Alves ALMM, dos Santos FV, dos Santos FA, Correa DS (2025) 3D-printed microneedles for sensing applications: emerging topics and future trends. Adv Sensor Energy Mater 4:100139. https://doi.org/10.1016/j.asems.2025.100139
2. Rai H, Singh KRB, Natarajan A, Pandey SS (2025) Advances in field effect transistor based electronic devices integrated with CMOS technology for biosensing. Talanta Open 11:100394
3. Hossain N, Mahmud MZA, Hossain A, Rahman MK, Islam MS, Tasnim R, Mobarak MH (2024) Advances of materials science in MEMS applications: a review. Results Eng 22:102115
4. Go G-e, Kim D (2025) Advancing biosensing through super-resolution fluorescence microscopy. Biosens Bioelectron 278:117374
5. Zhang Z, Zhou X, Fang Y, Xiong Z, Zhang T (2025) AI-driven 3D bioprinting for regenerative medicine: From bench to bedside. Bioactive Mater 45:201–230
6. Qureshi R, Irfan M, Gondal TM, Khan S, Wu J, Hadi MU, Heymach J, Le X, Yan H, Alam T (2023) AI in drug discovery and its clinical relevance. Heliyon 9:e17575
7. Shandhi MMH, Dunn JP (2022) AI in medicine: where are we now and where are we going? Cell Rep Med 3:100861
8. Ahmed AET, Dhahi TS, Attia TA, Ali FAE, Elobaid ME, Adam T, Gopinath SCB (2025) AI-optimized electrochemical aptasensors for stable, reproducible detection of neurodegenerative diseases, cancer, and coronavirus. Heliyon 11:e41338
9. Noushin T, Hossain NI, Tahrin R, Islam MN, Tabassum S (2025) An integrated kirigami-patterned skin patch for multiplexed detection of inflammatory biomarkers along with transdermal drug delivery. Sensing Bio-Sensing Res 47:100772
10. Karasu T, Çalışır F, Pişkin S, Özgür E, Armutcu C, Çorman ME, Uzun L (2024) An intriguing future is approaching: artificial intelligence meets molecularly imprinted polymers. J Pharm Biomed Anal Open 4:100041
11. Jin X, Liu C, Xu T, Su L, Zhang X (2020) Artificial intelligence biosensors: challenges and prospects. Biosens Bioelectron 165:112412
12. Lu M, Yin J, Zhu Q, Lin G, Mou M, Liu F, Pan Z, You N, Lian X, Li F, Zhang H, Zheng L, Zhang W, Zhang H, Shen Z, Gu Z, Li H, Zhu F (2023) Artificial intelligence in pharmaceutical sciences. Engineering 27:37–69

13. Qiu X, Zhu L, Wang H, Xi M (2025) Biocomputing at the crossroad between emulating artificial intelligence and cellular supremacy. Curr Opin Biotechnol 92:103264
14. Manickam P, Mariappan SA, Murugesan SM, Hansda S, Kaushik A, Shinde R, Thipperudraswamy SP (2022) Artificial intelligence (AI) and internet of medical things (IoMT) assisted biomedical Systems for Intelligent Healthcare. Biosensors 12:562. https://doi.org/10.3390/bios12080562
15. Atabay M, Inci F, Saylan Y (2025) Computational studies for the development of extracellular vesicle-based biosensors. Biosens Bioelectron 277:117275
16. Koh E, Sunil RS, Lam HYI, Mutwil M (2024) Confronting the data deluge: how artificial intelligence can be used in the study of plant stress, computational and structural. Biotechnol J 23:3454–3466
17. Kalambate PK, Kumar V, Dhanjai (2025) Decentralized electrochemical biosensors for biomedical applications: from lab to home. Next Nanotechnology 7:100128
18. Wang S, Wu M, Liu W, Liu J, Tian Y, Xiao K (2024) Dopamine detection and integration in neuromorphic devices for applications in artificial intelligence. Device 2:100284
19. Javaid M, Haleem A, Singh RP, Rab S, Suman R (2021) Exploring the potential of nanosensors: a brief overview. Sensors Int 2:100130
20. Jain S, Nehra M, Kumar R, Dilbaghi N, Hu T, Kumar S (2021) Ajeet Kaushik, Chen-zhong Li, internet of medical things (IoMT)-integrated biosensors for point-of-care testing of infectious diseases. Biosens Bioelectron 179:113074
21. Flynn CD, Chang D (2024) Artificial intelligence in point-of-care biosensing: challenges and opportunities. Diagnostics 14:1100
22. Nisa MU, Yaqub A, Khan MH, Yaseen F, Jilani S, Ajab H, Shah NS, Anazi AA (2025) Nanotechnology-driven electrochemical neurotransmitter sensing as a fundamental approach towards improving diagnostics and therapeutics: a review. Sensors Actuators Rep 9:100292
23. Paladhi AG, Manohar MV, Pal K, Vallinayagam S, Packirisamy ASB, Bashreer VA, Nandhini RS, Ukhurebor KE (2022) Novel electrochemical biosensor key significance of smart intelligence (IoMT & IoHT) of COVID-19 virus control management. Process Biochem 122:105–109
24. Rangel GF, Diaz L, Martínez d L´ o, Walter LS, Mizaikoff B (2024) Recent advances and trends in mid-infrared chem/bio sensors. Trends Anal Chem 180:117916
25. Sayyad PW, Park S-J, Ha T-J (2024) Recent advances in biosensors based on metal-oxide semiconductors system-integrated into bioelectronics. Biosens Bioelectron 259:116407
26. Zhou Y, Wang Q, Xiang T, Chen X (2025) Recent advances in electrochemical biosensors for bacterial detection. Nano Trans Med 4:100078
27. Chakraborty A, Dutta P, Wakankar A, RoyChaudhuri C (2022) Recent field effect transistors and electrical impedance spectroscopy based biosensing strategies for cancer biomarker screening: a mini review. Biosensors Bioelectron X 12:100253
28. Khan A, Al Rashid H, Kabiraz DC, Sarker AC, Chowdhury SI (2025) Self-powered wearable biosensors for metabolites and electrolytes detection: Harnessing nanogenerators and renewable energy sources. Sensors Actuators A Phys 386:116339
29. Boyd JD, Hoffman AF (2022) Advances in high-content imaging and informatics: a joint special collection with Society for Biomolecular Imaging and Informatics and SLAS. SLAS Discovery 26:1073–1075
30. Liang Y, Xiao R, Huang F, Lin Q, Guo J, Zeng W, Dong J (2024) AI nutritionist: intelligent software as the next generation pioneer of precision nutrition. Comput Biol Med 178:108711
31. Giannetto M, Bianchi V, Gentili S, Fortunati S, De Munari I, Careri M (2018) An integrated IoT-Wi-Fi board for remote data acquisition and sharing from innovative immunosensors. Case of study: diagnosis of celiac disease. Sensors Actuators B Chem 273:1395–1403
32. Nguyen HQ, Nguyen VD, Phan VM, Seo TS (2024) A novel point-of-care platform for rapid SARS-CoV-2 detection utilizing an all-in-one 3D-printed microfluidic cartridge and IoT technology. Sensors Actuators: B Chem 410:135632
33. Kobayashi H, Nakata N, Izuka S, Hongo K, Nishikawa M (2024) Using artificial intelligence and promoter-level transcriptome analysis to identify a biomarker as a possible prognostic

predictor of cardiac complications in male patients with Fabry disease. Mol Genet Metab Rep 41:101152

34. Cui R, Tang H, Huang Q, Ye T, Chen J, Huang Y, Hou C, Wang S, Ramadan S, Li B, Xu Y, Xu L, Li D (2024) AI-assisted smartphone-based colorimetric biosensor for visualized, rapid and sensitive detection of pathogenic bacteria. Biosens Bioelectron 259:116369
35. Dhandapani G, Wekalao J, Patel SK, Al-Zahrani FA (2024) Design and analysis of a highly sensitive terahertz biosensor using graphene metasurfaces and surface plasmon resonance for protein detection with AI-assisted locally weighted linear regression for behavior prediction. Plasmonics 20:3619. https://doi.org/10.1007/s11468-024-02563-3
36. Papa M, Wasit A, Pecora J, Bergholz TM, Yi J (2025) Detection of viable but Nonculturable E. coli induced by Low-level antimicrobials using AI-enabled hyperspectral microscopy. J Food Prot 88:100430
37. Choi JY, Park S, Shim JS, Park HJ, Kuh SU, Jeong Y, Park MG, Noh TI, Yoon SG, Park YM, Lee SJ, Kim H, Kang SH, Lee KH (2025) Explainable artificial intelligence-driven prostate cancer screening using exosomal multi-marker based dual-gate FET biosensor. Biosens Bioelectron 267:116773
38. Yang Y (2024) Application of wearable devices based on artificial intelligence sensors in sports human health monitoring. Measurement: Sensors 33:101086
39. Wang JH, Deek MP, Mendes AA, Song Y, Shetty A, Bazyar S, Van der Eecken K, Chen E, Showalter TN, Royce TJ, Todorovic T, Huei-Chung Huang SA, Houck R, Yamashita AP, Kiess DY, Song T, Lotan T, DeWeese L, Marchionni L, Ren PO (2025) Validation of an artificial intelligence-based prognostic biomarker in patients with oligometastatic castration-sensitive prostate cancer. Radiother Oncol 202:110618
40. Torok R, Meszaros B, Gombas V, Vathy-Fogarassy A, Szabo M, Csanky E, Jarvas G, Guttman A (2025) Predicting the effectiveness of chemotherapy treatment in lung cancer utilizing artificial intelligence-supported serum N-glycome analysis. Comput Biol and Med 186:109681
41. Park S, Kang H, Choi Y, Yoon SG, Park HJ, Jin H, Kim H, Jeong Y, Shim JS, Noh TI, Kang SH, Lee KH (2025) Precision screening with sequential multi-algorithm reclassification technique (SMART): saving bladders from unnecessary cystectomy. Comput Biol Med 189:109980
42. Low JSY, Teh HF, Thevarajah TM, Chang SW, Khor SM (2025) An AI-assisted microfluidic paper-based multiplexed surface-enhanced Raman scattering (SERS) biosensor with electrophoretic removal and electrical modulation for accurate acute myocardial infarction (AMI) diagnosis and prognosis. Biosens Bioelectron 270:116949
43. Dai F, Zhang T, Pang F, Jiao T, Wang K, Zhang Z, Wang N, Xie Z, Zhang Y, Wang Z, Chen Z, Yu M, Wei H, Song J (2025) A compact, palm-sized isothermal fluorescent diagnostic intelligent IoT device for personal health monitoring and beyond via one-tube/one-step LAMP-CRISPR assay. Biosens Bioelectron 270:116945
44. Wang Y, Tang F, Liu B, Wu Y, Zhang R, Ren H (2025) A bioinspired microbial taste chip with artificial intelligence-enabled high selectivity and ultra-short response time. Biosens Bioelectron 277:117264
45. Harnsoongnoen S, Srisai S, Kongkeaw P, Buranrat B (2025) Detection and classification of glucose solution concentration and blood sugar levels at the fingertip using a novel planar microwave sensor and deep learning techniques. Sensors Actuators: B Chem 430:137322
46. Zhao J, Guo F, Wang M, Zhang J, Ying S, Gao Y, Yang G, Hong W (2025) The advancement of biosensor design and construction utilizing biomolecular motors, synthetic and systems. Biotechnology 10:543–554
47. Träger TK, Tüting C, Kastritis PL (2024) The human touch: utilizing alpha fold 3 to analyze structures of endogenous metabolons. Structure 32:1555–1562
48. Ronish LA, Biswas B, Bauer RM, Jacob ME, Piepenbrink KH (2024) The role of extracellular structures in Clostridioides difficile biofilm formation. Anaerobe 88:102873
49. Kutashev K, Moschou PN (2025) The role of membrane-biomolecular condensate interactions in stress. Plant Stress 16:100810

50. Adil O, Shamsi MH (2025) Transformative biomedical devices to overcome biomatrix effects. Biosens Bioelectron 279:117373
51. Nerella S, Bandyopadhyay S, Zhang J, Contreras M, Siegel S, Bumin A, Silva B, Sena J, Shickel B, Bihorac A, Khezeli K, Rashidi P (2024) Transformers and large language models in healthcare: a review. Artif Intell Med 154:102900
52. Khera R, Oikonomou EK, Nadkarni GN, Morley JR, Wiens J, Butte AJ, Topol EJ (2024) Transforming cardiovascular care with Artificial Intelligence: From discovery to practice. JACC 84:97–114
53. Islam MR, Khan MMI, Al Rafid R, Mehjabin F, Shahriar Rashid M, Chowdhury JA, Zerin N, Islam M (2022) Trigonal cluster-based ultra-sensitive surface plasmon resonance sensor for multipurpose sensing. Sensing Bio-Sensing Res 35:100477
54. Park JY, Kwak Y, Lim H-R, Si-Woo Park MS, Lim H-BC, Myung NV, Choa Y-H (2022) Tuning the sensing responses towards room-temperature hypersensitive methanol gas sensor using exfoliated graphene-enhanced ZnO quantum dot nanostructures. J Hazard Mater 438:129412
55. Fuellen G, Kulaga A, Lobentanzer S, Unfried M, Avelar RA, Palmer D, Kennedy BK (2025) Validation requirements for AI-based intervention-evaluation in aging and longevity research and practice. Ageing Res Rev 104:102617
56. Kim J, McFee M, Fang Q, Abdin O, Kim PM (2023) Computational and artificial intelligence-based methods for antibody development. Trends Pharmacol Sci 44:175–189
57. Kumar M, Ashok AK, Bhat T, Ballamoole K, Gollapalli P (2025) Computational-driven discovery of AI-2 quorum sensing inhibitor targeting the 5′-methylthioadenosine/S--adenosylhomocysteine nucleosidase (MTAN) to combat drug-resistant Helicobacter pylori. Comput Biol Med 184:109409
58. Gentili PL, Stano P (2024) Living cells and biological mechanisms as prototypes for developing chemical artificial intelligence. Biochem Biophys Res Commun 720:150060
59. Navarro J, Marijuan PC (2023) Natural intelligence and the ‘economy’ of social emotions: a connection with AI sentiment analysis. Biosystems 233:105039
60. Mohanty SP, Srivastava A, Hu S, Ghosal P (2017) Guest editorial—special issue on hardware assisted techniques for IoT and bigdata applications. Integration 58:263–266

Chapter 9
Eco-Friendly Synthesis of Nanomaterials for Pesticide Sensing

Jayanti Mishra, Priya Ghosh, Himangshu Pratim Bhattacharyya, and Arabinda Baruah

Abstract This chapter discusses about the green synthesis of various nanomaterials by plants, biomolecules, and microorganisms for the application in pesticides recognition. Plants parts mediate the synthesis of metal nanoparticles by producing phytochemicals which act as natural reducing and capping agents. Neem, aloe-vera, green tea, tulsi, hibiscus, lemon peel, pomegranate peel, etc., are the popular plants used for nanoparticle synthesis via green route. This chapter also discusses about the synthesis of nanomaterials like metal, metal oxides, metal sulfides, chalcogenides, etc., facilitated by bacteria and fungi for application in detection of harmful pesticides. Many biomolecules like enzymes, proteins, NADH, DNA, aptamers, etc., are reported for synthesis of nanomaterials which have shown the remarkable recognition of pesticides. The scope of this chapter is to discusses the role of plants, biomolecules, microorganisms, and algae/sea weeds toward synthesis of variety of nanomaterials and their application in the biosensing of pesticides like thiram, malathion, glycophosphate, triazophos, etc.

Keywords Nanomaterial · Pesticides · Plants · Microorganisms · Biomolecules · Sensing

---

J. Mishra (✉)
Department of Applied Sciences, New Horizon College of Engineering, Bangalore, Karnataka, India

P. Ghosh
Department of Chemistry, GLC College, Barpeta Road, Assam, India

H. P. Bhattacharyya
Department of Chemistry, Indian Institute of Technology Guwahati, Guwahati, Assam, India

Department of Chemistry, Gauhati University, Guwahati, Assam, India

A. Baruah
Department of Chemistry, Gauhati University, Guwahati, Assam, India

R. Naik et al. (eds.), *Biocompatible Nanomaterials for Smart Sensors*, Nanostructure Science and Technology,
https://doi.org/10.1007/978-3-032-01745-1_9

9.1 Introduction

To protect crops from pests and insects and enhance their yield, farmers around the world rely on pesticides. In 2022, approximately 2.66 million tons of pesticides were used globally [1]. Although the chemicals present in the pesticide are targeted for the harmful pests, only a fraction of chemicals reach the target. The remaining chemicals get deposited on the surface of the soil and affect the quality of soil and water. As many pesticides persist long in the environment, the chemicals cause gradual but significant harm to the ecosystem. Therefore, it is highly essential to develop techniques to detect pesticides. However, for detection, the development of biosensors is one of the viable alternatives. Three categories of pesticides primarily raised concerns about the environment. Among them, organochlorines, organophosphates, and carbamates are well-known examples. In addition, organochlorines are also a harmful category of pesticides that utilize chemicals, out of which DDT is the most famous example [2].

Nanomaterials are the materials whose at least one dimension lies in the range of 1–100 nm. They are becoming essential components in the development of biosensors due to their ability to significantly improve sensor performance. Due to their tunable size and morphology, they exhibit different properties as compared to their bulk counterparts. By tailoring the size, morphology, surface area, surface defects, and band gap, the electronic, optical, electrical, thermal, magnetic, and mechanical properties can be tuned easily. Major areas of utilization of nanomaterials lie in electronics, optics, therapeutics, drug delivery, dentistry, laser instruments, piezoelectrics, catalytics, and sensing. Their unique optical and electrical properties, high surface-to-volume ratios, excellent conductivity, and catalytic capabilities, enable faster, more efficient, and cost-effective detection. Their compatibility with biological systems and the ease with which they can be functionalized, they have vast application for specific sensing applications like chemosensing, colorimetric sensing, potentiometric sensing, fluorescent sensing, etc. [3] Common synthesis approaches for synthesis of nanoparticles are sol-gel, hydrothermal, colloidal and microemulsion methods [4, 5]. Surface functionalization during or after synthesis, tunes the properties of nanomaterials as per application in sensing and catalysis. Surface functionalization of ZnO by urea based imidazole moiety and Schiff base moiety have shown the degradation of Rhodamine B dye to a great extent [6–8]. Organic nanoparticles of urea containing Schiff base tetrapodal compound has shown the chemosensing of Pb(II) in aqueous medium.

Out of various ways to synthesize nanomaterials, precursors obtained from biological route gives an edge because of being environmentally friendly thus being sustainable [9]. Here, in this chapter, we are discussing the green synthesis of nanomaterials from sources including plants, microorganisms, biomolecules, sea weeds, algae, mushrooms, etc. This is making the synthesis procedure environment friendly, economic and more effective.

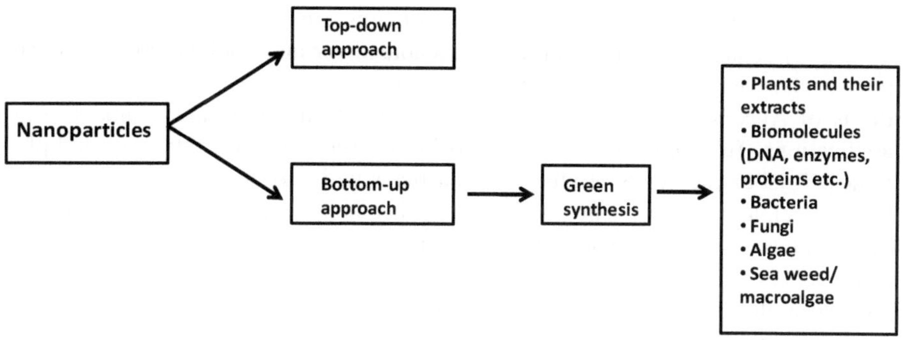

Fig. 9.1 Green synthesis approach of nanoparticle synthesis

9.2 Green Synthesis of Nanomaterials

Figure 9.1 shows the green synthesis approach for nanoparticle synthesis is environment friendly, cost effective, and easy to handle.

9.2.1 Plant-Extract-Mediated Preparation of Nanoparticles for Pesticide Detection

Numerous methodologies have been developed for the synthesis of nanomaterials [10–14]. Among these, plant extract-mediated synthesis has emerged as a widely adopted green and sustainable approach. This method leverages the inherent phytochemicals present in plant materials, such as flavonoids, alkaloids, terpenoids, and phenolic compounds, which act as natural reducing and capping agents to facilitate the conversion of metal ions into nanoparticles. The popularity of this route stems from its operational simplicity, environmental compatibility, and cost-effectiveness. Furthermore, in the context of pesticide sensing, it is essential that the synthesized nanoparticles exhibit surface functionalities capable of specific interactions with target analytes. Plant-based synthesis inherently introduces a diverse array of organic moieties onto the nanoparticle surface, thereby enhancing their chemical affinity and selectivity toward pesticides, and making this approach particularly advantageous for sensor development. Various bioactive compounds present in plants, such as flavonoids, terpenoids, alkaloids, phenolic acids, saponins, and proteins, act as both reducing and stabilizing agents in the synthesis of nanoparticles.

Owing to their advantageous features, in the recent years, plant-mediated (or phyto-mediated) synthesis of nanoparticles has emerged as a highly popular field of research. Among numerous plant sources used, neem, green tea, aloe-vera, tulsi, hibiscus, lemon peel, and pomegranate peel are prominent. *Azadirachta indica* or neem is rich in flavonoids and terpenoids. The neem extract has been extensively used to synthesize silver, gold, and copper nanoparticles with strong antimicrobial properties. Green tea (*Camellia sinensis*) contains catechins and polyphenols that effectively reduce metal ions, producing stable silver and gold nanoparticles. The polysaccharides and phenolic compounds in *Aloe vera* extract facilitate rapid synthesis of AgNPs and ZnO NPs, often used in antimicrobial and wound healing applications. Tulsi or *Ocimum sanctum* is known for its rich essential oils and phenolics. Tulsi extract helps produce nanoparticles with bioactive surfaces. Hibiscus, lemon peel, and pomegranate peel have also been employed, contributing a mix of antioxidants and organic acids for efficient nanoparticle formation. The advantages of plant-based synthesis include low toxicity, cost-effectiveness, biocompatibility, and the ability to tailor nanoparticle properties through plant choice and extract concentration. These nanoparticles find applications in diverse domains of research. As research progresses, more exotic and underutilized plants are being explored, unlocking new possibilities for greener nanotechnology.

Typically, to obtain the necessary plant extracts for nanoparticle synthesis, selected plant parts, such as leaves, flowers, roots, stems, or fruits, are thoroughly washed, dried, and subjected to boiling in water or ethanol to extract bioactive phytochemicals. The resulting extract is then filtered and stored under appropriate conditions to prevent physical, chemical, or biological contamination. For nanoparticle synthesis, metal salts are generally reacted with the prepared plant extracts under controlled temperature and pH conditions. In some cases, additional additives are incorporated to enhance the nanoparticles' affinity for pesticide molecules. In the subsequent sections, recent studies on pesticide detection utilizing nanoparticles synthesized through green chemical approaches are reviewed. Emphasis is placed on the mechanistic aspects of nanoparticle formation and their interaction with pesticide molecules to offer a comprehensive understanding of this eco-friendly sensing strategy.

In a fascinating report by Alam et al. gold and silver nanoparticles have been synthesized via a green, bio-friendly method using plant extracts, specifically the aqueous methanol extract of Indian curry leaf (*Murraya koenigii*) leaves as a green multifunctional agent [15]. The extract is rich in bioactive compounds, including carbazole alkaloids and polyhydroxy flavonoids (e.g., quercetin), which act as natural reducing and stabilizing agents. Under sonochemical conditions, the plant extract enabled the rapid and controlled synthesis of monometallic silver (Ag) nanoparticles, monometallic gold (Au) nanoparticles and bimetallic Ag/Au alloy nanoparticles. These nanoparticles were stable, crystalline, and fluorescent, with unique morphologies influenced by the natural compounds in the extract. The synthesized nanoparticles demonstrated strong fluorescence, which was leveraged for the fluorometric detection of mancozeb, a widely used dithiocarbamate pesticide. Upon addition of mancozeb to the nanoparticle solution, there was a "fluorescence turn-on" effect, significantly increasing fluorescence intensity. This is attributed to the interaction between mancozeb and nanoparticle surfaces, which displaces surface-bound fluorescent plant molecules, thereby increasing their concentration in solution and enhancing the signal. This method showed sensitivity in aqueous medium, highlighting the nanoparticles' potential as eco-friendly pesticide sensors.

Linh et al [16]. reported an eco-friendly electrochemical method to make silver nanoparticles (AgNPs) using three coating agents: citrate (e-Ag-C), oleic acid (e-Ag-O), and a plant-based fish mint extract (e-Ag-bio). These nanoparticles were tested as SERS (surface-enhanced Raman spectroscopy) substrates to detect very small amounts of the pesticide thiram. Among the three, e-Ag-O worked best because its particles were closer together (~4 nm apart), creating stronger "hotspots" that boosted the Raman signal. e-Ag-C adsorbed more thiram, which helped chemical enhancement, but its larger spacing (~7 nm) reduced overall SERS intensity compared to e-Ag-O. The green-synthesized e-Ag-bio had the lowest crystallinity and interference from plant chemicals, yet still performed as well as e-Ag-C at higher thiram concentrations. However, it was less effective at very low concentrations, limiting its use for ultra-trace detection. The e-Ag-O substrate achieved a detection limit of 1.04×10^{-10} M and worked well across two linear ranges. It also successfully detected thiram in green tea leaves with recovery rates of 93–132%, proving its real-world potential. Although e-Ag-bio was not the top performer, the study shows that plant-based green synthesis can produce sustainable, eco-friendly SERS sensors.

9.2.2 *Microorganism-Assisted Synthesis of Nanomaterials for Pesticide Sensing*

Microbial synthesis of nanomaterials offers a sustainable and eco-friendly alternative to conventional chemical and physical methods [17]. Certain microorganisms, especially bacteria, have evolved the ability to reduce toxic metal and metalloid ions into stable nanoparticles (NPs). This process, which can occur intracellularly or extracellularly, leverages microbial enzymes, metabolites, and cellular structures to facilitate nanoparticle formation. Intracellular synthesis involves ion uptake and reduction within the cell using intracellular enzymes and cofactors. Extracellular synthesis uses secreted enzymes or cell wall components to reduce ions in the surrounding medium.

Microorganisms have been used to synthesize a variety of metal, metalloid, and compound nanoparticles, including Ag, Au, Cu, Pd, Se, and Te; metal oxides (Fe_3O_4, SnO_2); sulfides (PbS, ZnS) and chalcogenides (CdS, CdTe); etc. [18]. These microbially synthesized NPs have shown great potential in various applications, such as, environmental remediation (e.g., pesticide detection, degradation of dyes like methylene blue and Rhodamine B) dehalogenation of toxic chlorinated compounds, reduction of nitroaromatics (e.g., 4-nitrophenol) heavy metal removal from wastewater as well as hydrogen generation and fuel cell reactions [18]. These NPs are often more biocompatible, stable, and catalytically efficient than those made via conventional methods, primarily due to their high surface area and biologically derived coatings.

Bacterial species such as *Bacillus*, *Stenotrophomonas*, *Pseudomonas*, *Deinococcus*, *Shewanella*, *Lactobacillus*, and *Escherichia coli* have been successfully employed to generate silver, gold, copper, and palladium nanoparticles. Additionally, organisms like *Rhodococcus*, *Citrobacter*, and *Enterobacter* are known for synthesizing selenium and tellurium nanoparticles, while *Magnetospirillum* and *Erwinia* contribute to the formation of magnetic and semiconducting metal oxides such as Fe_3O_4 and SnO_2. These nanoparticles exist in monometallic, bimetallic, oxide, and sulfide forms, each offering unique properties for applications in catalysis, environmental remediation, and biosensing. Their biocompatibility, high surface reactivity, and structural diversity position them as versatile tools in sustainable technology development. In the following section, we delve into the promising applications of these biologically synthesized nanoparticles for pesticide detection, citing key advancements from recent literature.

In literature, there are numerous reports on biogenic nanoparticles for pesticide detection and removal [19]. However, reports on pesticide sensing using microbially synthesized nanoparticles are very few. For instance, fungi such as *Fusarium*, *Verticillium*, *Penicillium*, and *Aspergillus* have been utilized to synthesize various metal nanoparticles, including silver and gold. These biosynthesized nanoparticles have demonstrated significant potential in environmental remediation and sensing applications. For example, silver nanoparticles synthesized using *Rhizopus oryzae* have been explored for their ecological applications, including wastewater treatment and adsorption of pesticides [20]. Biosynthesized gold nanoparticles (AuNPs) have also been employed in the development of chemical sensors for detecting organochlorine pesticides. These sensors leverage the unique optical properties of AuNPs to achieve sensitive and selective detection of pesticide residues. Goel *et. al* [21]. have reported that gold nanoparticles serve as effective colorimetric sensors for detecting organochlorine pesticides like endosulfan due to their unique optical properties. In the study, gold nanoparticles of nearly 8 nm size were utilized to detect endosulfan pesticide (ESP). The presence of ESP induces aggregation of gold nanoparticles, leading to a visible color change from wine red to various shades of blue. This color shift results from alterations in the surface plasmon resonance of the nanoparticles, which is sensitive to changes in their aggregation state. Additionally, a gold nanoparticles-based sensing electrode was developed for quantitative detection of ESP at ambient temperature. The sensor's response and sensitivity were evaluated by monitoring changes in electrical resistance over time, providing a reliable method for pesticide detection. This approach offers a simple, rapid, and cost-effective means for monitoring pesticide residues in environmental samples..

Mirghafouri et al. have reported nanoplasmonic sensor arrays, fabricated using biosynthesized nanoparticles, for the detection and discrimination of pesticide residues in agricultural products like citrus fruits. These sensors utilize the plasmonic properties of nanoparticles to achieve rapid and sensitive detection [22]. In this study, a non-enzymatic sensor array was designed to detect and distinguish five different pesticides from various chemical classes, including organophosphates, carbamates, and bipyridyliums. Two types of gold nanoparticles, capped with citrate and borohydride, respectively, served as the sensing elements. The detection relied on changes in the spectral properties of gold nanoparticle induced by their aggregation, enabling pesticide identification across a broad concentration range (20–5000 ng mL^{-1}).

In a review by Bahrulolum et al. (2021) the green synthesis of metal nanoparticles (MtNPs) using microorganisms and their transformative applications in the agrifood sector have been thoroughly discussed. Microorganisms such as bacteria, fungi, yeasts, and microalgae act as eco-friendly nanofactories, capable of reducing metal ions into nanoparticles via enzymatic and metabolic processes. These biosynthesized nanoparticles, including Ag, Au, Cu, Zn, and Se, exhibit antimicrobial, antifungal, and plant growth-promoting properties, making them highly valuable in agriculture. The review highlights the use of metal NPs in nanopesticides, nanofungicides, nanofertilizers, and nanobiosensors, offering controlled release, improved efficacy, and reduced environmental toxicity compared to conventional agrochemicals. Intracellular and extracellular biosynthesis mechanisms are discussed, emphasizing the role of enzymes like nitrate reductase and NADH-dependent reductases. Here, the authors have also described probiotic- and non-probiotic bacterial strains (e.g., *Lactobacillus*, *Escherichia coli*, *Pseudomonas*) and fungal species (e.g., *Fusarium*, *Aspergillus*) used for metal NP synthesis [23].

Thus, it can be concluded that microorganism-assisted synthesis of nanoparticles represents a highly sustainable and eco-conscious approach to nanomaterial production. This method not only eliminates the need for toxic chemicals and energy-intensive processes typically used in conventional synthesis, but also contributes to environmental monitoring by sensing pesticides and detoxification by enabling microorganisms to sequester and transform hazardous metal ions into benign, functional nanostructures. Acting as natural nanofactories, a wide variety of microorganisms, including bacteria, fungi, and microalgae, have demonstrated remarkable capabilities in the biosynthesis of diverse nanomaterials.

9.2.3 Biomolecules-Mediated Synthesis of Nanoparticles for Pesticide Sensing

Biomolecules like enzymes, DNA, aptamers are known for synthesis of Au/Ag nanoparticles, which has shown effective sensing of pesticides like chlorpyrifos, malathion, paraoxon, etc.

Among different enzyme-mediated nanomaterials, the synthesis of gold nanomaterials (AuNMs) exhibits distinct optical and electronic properties due to their strong surface plasmon resonance (SPR) in the visible range. The interaction of electromagnetic radiation with the AuNM can be used to enhance the sensitivity for pesticide detection. For instance, AuNMs are used to develop biosensors to detect methomyl with a limit of detection (LOD) of 81 ng/L. Further, AuNMs have been used to detect the toxic organophosphate chlorpyrifos in vegetables. To an aptamer, the ferrocyanide probe is attached and fixed on the electrode surface using gold nanomaterials and molybdenum compounds. This forms a stable double-stranded structure. When chlorpyrifos binds to the aptamer, the structure breaks, and the ferrocyanide moves closer to the electrode, enhancing the signal [24].

Similarly, due to their large surface area and fast electron movement, the silver nanomaterials (AgNMs) are considered another suitable choice for pesticide sensing. Wang and Liu developed a fluorescent biosensor using guanine-rich DNA and silver nanoparticles (AgNPs) doped with Terbium (Tb^{3+}) to detect organophosphorus pesticides in apples. When exposed to UV light, energy transfers from the DNA to Tb^{3+}, enhancing fluorescence and enabling detection as low as 34 ng/L. [25]. In another study, Bala et al. have created a color-based biosensor using AgNPs and malathion-specific aptamers. The biosensor showed the color change from yellow to orange in the presence of malathion [26]. In addition, the AgNPs showed their utility in electrochemical biosensors for detecting paraoxon in vegetables [27]. Along the line, silver nanowires (AgNWs) have also been incorporated into polymer-coated electrodes for detecting paraoxon in milk. These examples illustrate the cost-effectiveness and good sensitivity offered by AgNMs in sensing pesticides.

Wang et al. developed an electrochemical biosensor using a graphene membrane with magnetic nanoparticles (Fe_3O_4) and acetylcholinesterase (AChE) inhibition to detect chlorpyrifos. The membrane has a large specific surface area and high electron transfer. This allows effective immobilization of AChE and better detection. The nanocomposite provides many active sites that support catalysis reactions. It also improves the AChE reaction. The biosensor showed excellent sensitivity with a LOD of 20 ng/L [28]. In the presence of the pesticide, the aptamer binds to it, separating from the nanozyme and triggering a color reaction. In addition, Dong et al. developed an electrochemical biosensor for detecting malathion and methyl parathion in Chinese cabbage. The detection was based on the inhibition of AChE using a three-dimensional film of rGO with nickel foam. It was combined with AuNPs to form AuNPs/rGO/NF, which provides a large surface area for AChE adsorption. It also ensures excellent electron transfer due to the synergistic effect of AuNPs and rGO, and shows sensitivity LOD of 0.0217 ng/L [29].

Yang et al. developed a biosensor using Ir nanorods with Cadmium Sulfide Quantum Dots (Ir NRs@CdSQDs) to detect paraoxon in cabbage pakchoi and lettuce [30]. The biosensor uses electrochemiluminescence, in which Ir NRs act as an anodic emitter and CdS QDs act as a cathodic emitter. In the presence of organophosphate pesticides, a bio-composite of AChE and choline oxidase (ChO_x) is immobilized and produces H_2O_2. This increases the cathode signal and decreases the anodic signal. Thus, the biosensor has long-term stability [30]. Moreover, Korram et al. developed a device with CdTe QDs that allows sensitive detection without spectral interference. The fluorescent biosensor is based on AChE inhibition that detects paraoxon, dichlorvos, malathion, and triazophos in apple and tomato juice. The method mixes the sample with AChE and a solution containing AChE, ChO_x, and QDs and is more sensitive to paraoxon and less to malathion [31]. The magnetic Fe_3O_4 nanoparticles are bound to a chlorpyrifos-specific aptamer, and the CuO-metal organic framework enzyme-mediated nanomaterials are linked to complementary DNA strands [32].

9.2.4 Algae and Sea Weeds-Mediated Synthesis of Nanoparticles for Pesticide Sensing

Algae are less toxic, applicable for ambient temperature and easy to handle for synthesis of nanoparticles a normal pH value. Phyco-nanotechnology term is given to the branch under which algae-mediated synthesis of metal nanoparticles are studied. However, application of synthesized nanoparticles in pesticide sensing is still under discussion. Sea weed are also called as macro algae. Metal nanoparticles synthesized by algae and sea weed extract are synthesized by rapid and non-toxic process. Shanmugam et al. have reported biological synthesis of Ag nanoparticles by the reduction of silver nitrate to aqueous silver metal ions using marine seaweed *Sargassum wightii* extract [33]. Rajeshkumar et al. synthesized Au nanoparticles (6–10 nm) by green route using algae extract of *Turbinaria conoides* which has shown biomedical applications [34]. Cyanobacteria (blue-green algae) is known for its role in nanoparticle synthesis. Whole cells of *Plectonema boryanum*, *Anabaena flos-aquae*, and *Calothrix pulvinate* are reported for production of Au, Ag, and Pt nanoparticles. Apart from them, *Lyngbya majuscule* are *Chlorella vulgaris* are reported for Ag nanoparticle synthesis. *Spirulina platensis* (dried edible algae) are known for extracellular synthesis of Au, Ag, and Au/Ag bimetallic nanoparticles. *Fucus vesiculosus* (brown alga) is used for bioreduction of Au(III) to Au(0). *Chondrus crispus* (red alga) and *Spyrogira insignis* (green alga) are also known for Au nanoparticles synthesis [35].

9.3 Conclusion

This chapter discussed the green synthesis of tailored nanomaterials by plants, biomolecules and microorganisms. Plants parts such as leaves, flowers, roots, stems, or fruits are known for the source of various chemicals like alkaloids, flavonoids, terpenoids, phenolic compounds, saponins, catechins, polyphenols, polysaccharides, etc. These natural products facilitate the synthesis of nanomaterials by working as surface directing agents. Plant parts of common plants like neem, aloe-vera, green tea, tulsi, hibiscus, lemon peel, curry leaves etc., are reported for synthesis of Ag, Au, Cu, and ZnO nanoparticles etc. Bacteria and fungi are common microorganisms which help to produce nanomaterials. Few bacteria such as *Bacillus*, *Stenotrophomonas*, *Pseudomonas*, *Deinococcus*, *Shewanella*, *Lactobacillus*, *Escherichia coli, Rhodococcus*, *Citrobacter*, *Enterobacter*, *Magnetospirillum*, *Erwinia* etc., and fungi such as *Fusarium*, *Rhizopus*, *Verticillium*, *Penicillium*, and

Aspergillus etc., are known for the synthesis of nanoparticles of Ag, Au, Cu, Pd, Se, Te, Fe_3O_4, SnO_2, PbS, ZnS, CdS, CdTe etc. Biomolecules such as DNA, aptamers, enzymes etc., are known for the synthesis of metal nanoparticles, which have shown effective sensing of pesticides like chlorpyrifos, malathion, paraoxon, etc. Apart from plant parts and microorganisms mentioned above, seaweeds and algae are reported for the synthesis of Ag, Au and Pt nanomaterials remarkably as well. The resulting compounds are being used for sensing of various pesticides like dithiocarbamate-based mancozeb, thiram, malathion, glycophosphate, triazophos, acetamiprid, chlorpyrifos, endosulfan, etc.

References

1. Zhang C, Qiu M, Wang J, Liu Y (2023) Recent advances in nanoparticle-based optical sensors for detection of pesticide residues in soil. Biosensors 13(4):415. https://doi.org/10.3390/bios13040415
2. Gansmüller A, Concistrè M, McLean N, Johannessen OG, Marín-Montesinos I, Bovee-Geurts PH, Verdegem P, Lugtenburg J, Brown RC, DeGrip WJ, Levitt MH (2009) Towards an interpretation of 13C chemical shifts in bathorhodopsin, a functional intermediate of a G-protein coupled receptor. Biochim Biophys Acta (BBA)—Biomem 1788(6):1350–1357. https://doi.org/10.1016/j.bbamem.2009.02.018
3. Aragay G, Pino F, Merkoçi A (2012) Nanomaterials for sensing and destroying pesticides. Chem Rev 112(10):5317–5338. https://doi.org/10.1021/cr300020c
4. Logothetidis S (2011) Nanotechnology: principles and applications. In: Nanoscience and technology, pp 1–22. https://doi.org/10.1007/978-3-642-22227-6_1
5. Mekuye B, Abera B (2023) Nanomaterials: an overview of synthesis, classification, characterization, and applications. Nano Select 4(8):486–501. https://doi.org/10.1002/nano.202300038
6. Mishra J, Kaur N, Ganguli AK (2018) Selective and sensitive fluorescence recognition of Pb(II) in aqueous medium by organic nanoparticles of a urea linker based tetrapodal receptor: effect of linker molecules in a sensor on chemosensing. Inorg Chim Acta 487:214–220. https://doi.org/10.1016/j.ica.2018.12.022
7. Mishra J, Jha M, Kaur N, Ganguli AK (2018) Room temperature synthesis of urea based imidazole functionalised ZnO nanorods and their photocatalytic application. Mater Res Bull 102:311–318. https://doi.org/10.1016/j.materresbull.2018.02.045
8. Mishra J, Kaur R, Kaur N, Ganguli AK (2022) Surface functionalization of ZnO nanoparticles by Schiff base oriented tetrapodal ligands at room temperature and their photocatalytic applications: a comparative account on effect of structure directing agent on photocatalysis. Inorg Chem Commun 146:110130. https://doi.org/10.1016/j.inoche.2022.110130
9. Sanket S, Das SK (2021) Role of enzymes in synthesis of nanoparticles. Springer, pp 139–153. https://doi.org/10.1007/978-981-33-4195-1_7
10. Bressi V, Ferlazzo A, Iannazzo D, Espro C (2021) Graphene quantum dots by eco-friendly green synthesis for electrochemical sensing: recent advances and future perspectives. Nano 11(5):1120. https://doi.org/10.3390/nano11051120
11. Newar R, Sultana N, Das S, Gogoi B, Islam NF, Sarma H, Baruah A (2023) Development of FRET-based optical sensors using N-doped carbon dots for detection of chromium (VI) and manganese (VII) in water for a sustainable future. J Environ Chem Eng 12(1):111721. https://doi.org/10.1016/j.jece.2023.111721
12. Mishra J, Baruah A, Dutta J (2024) Advanced synthesis strategies for enhanced energy storage performance. In: Nanostructure science and technology, pp 99–110. https://doi.org/10.1007/978-3-031-74730-4_6

13. Mishra J, Baruah A (2024) ZnO-based nanostructured materials for photocatalytic applications. CRC Press, pp 313–324. https://doi.org/10.1201/9781003479239-20
14. Baruah A, Newar R, Das S, Kalita N, Nath M, Ghosh P, Chinnam S, Sarma H, Narayan M (2024) Biomedical applications of graphene-based nanomaterials: recent progress, challenges, and prospects in highly sensitive biosensors. Discover Nano 19(1):103. https://doi.org/10.1186/s11671-024-04032-6
15. Alam MN, Das S, Batuta S, Mandal D, Begum NA (2016) Green-nanochemistry for safe environment: bio-friendly synthesis of fluorescent monometallic (Ag and Au) and bimetallic (Ag/Au alloy) nanoparticles having pesticide sensing activity. J Nanostructure Chem 6(4):373–395. https://doi.org/10.1007/s40097-016-0209-y
16. Linh DT, Mai Q, Nga DTN, Anh NT, Van Tuan H, Nguyen HA, Vu XH, Le A (2024) Surface ligand modified silver nanoparticles-based SERS sensing platform for ultrasensitive detection of the pesticide thiram in green tea leaves: roles of coating agents in sensing performance. RSC Adv 14(14):9975–9984. https://doi.org/10.1039/d4ra00048j
17. Tauseef A, Hisam F, Hussain T, Caruso A, Hussain K, Châtel A, Chénais B (2022) Nanomicrobiology: emerging trends in microbial synthesis of nanomaterials and their applications. J Clust Sci 34(2):639–664. https://doi.org/10.1007/s10876-022-02256-z
18. Fang X, Wang Y, Wang Z, Jiang Z, Dong M (2019) Microorganism assisted synthesized nanoparticles for catalytic applications. Energies 12(1):190. https://doi.org/10.3390/en12010190
19. Rani M, Anoushka N, Hosseini-Bandegharaei A, Shanker U (2024) Pesticides removal and their detection in real samples using green synthesized nanocomposites of biogenic quantum dots and metal oxides: a comprehensive review on recent updates. J Mol Liq 413:125921. https://doi.org/10.1016/j.molliq.2024.125921
20. Tazeen SK, Maqsood A, Asim Z, Mahmood F, Shahid M (2025) Nano-bioremediation of pesticides: an avenue to environmental safety. In: Smart waste and wastewater management by biotechnological approaches, pp 179–192. https://doi.org/10.1007/978-981-97-8673-2_9
21. Goel P, Arora M (2018) Fabrication of chemical sensor for organochlorine pesticide detection using colloidal gold nanoparticles. MRS Commun 8(3):1000–1007. https://doi.org/10.1557/mrc.2018.125
22. Mirghafouri MR, Abbasi-Moayed S, Ghasemi F, Hormozi-Nezhad MR (2020) Nanoplasmonic sensor array for the detection and discrimination of pesticide residues in citrus fruits. Anal Methods 12(48):5877–5884. https://doi.org/10.1039/d0ay02039g
23. Mandal M, Sarkar A (2024) Green syntheses of nanoparticles from plant growth–promoting microorganisms and their application in the agri-food industries. Elsevier, pp 185–204. https://doi.org/10.1016/b978-0-323-99682-2.00010-4
24. Lin Z, Liu X, Li Y, Li C, Yang L, Ma K, Zhang Z, Huang H (2021) Electrochemical aptasensor based on Mo2C/Mo2N and gold nanoparticles for determination of chlorpyrifos. Microchim Acta 188(5):170. https://doi.org/10.1007/s00604-021-04830-0
25. Wang T, Liu Y (2021) A lanthanide-based ratiometric fluorescent biosensor for the enzyme-free detection of organophosphorus pesticides. Anal Methods 13(17):2005–2010. https://doi.org/10.1039/d1ay00345c
26. Bala R, Mittal S, Sharma RK, Wangoo N (2018) A supersensitive silver nanoprobe based aptasensor for low cost detection of malathion residues in water and food samples. Spectrochimica Acta Part A Mol Biomol Spectroscopy 196:268–273. https://doi.org/10.1016/j.saa.2018.02.007
27. Turan J, Kesik M, Soylemez S, Goker S, Coskun S, Unalan HE, Toppare L (2016) An effective surface design based on a conjugated polymer and silver nanowires for the detection of paraoxon in tap water and milk. Sensors Actuators B Chem 228:278–286. https://doi.org/10.1016/j.snb.2016.01.034
28. Wang H, Zhao G, Chen D, Wang Z, Liu G (2016) A sensitive acetylcholinesterase biosensor based on screen printed electrode modified with Fe_3O_4 nanoparticle and graphene for Chlorpyrifos determination. Int J Electrochem Sci 11(12):10906–10918. https://doi.org/10.20964/2016.12.90

29. Dong P, Jiang B, Zheng J (2019) A novel acetylcholinesterase biosensor based on gold nanoparticles obtained by electroless plating on three-dimensional graphene for detecting organophosphorus pesticides in water and vegetable samples. Anal Methods 11(18):2428–2434. https://doi.org/10.1039/c9ay00549h
30. Yang G, He Y, Zhao J, Chen S, Yuan R (2021) Ratiometric electrochemiluminescence biosensor based on Ir nanorods and CdS quantum dots for the detection of organophosphorus pesticides. Sensors Actuators B Chemical 341:130008. https://doi.org/10.1016/j.snb.2021.130008
31. Korram J, Dewangan L, Karbhal I, Nagwanshi R, Vaishanav SK, Ghosh KK, Satnami ML (2020) CdTe QD-based inhibition and reactivation assay of acetylcholinesterase for the detection of organophosphorus pesticides. RSC Adv 10(41):24190–24202. https://doi.org/10.1039/d0ra03055d
32. Liu Q, He Z, Wang H, Feng X, Han P (2020) Magnetically controlled colorimetric aptasensor for chlorpyrifos based on copper-based metal-organic framework nanoparticles with peroxidase mimetic property. Microchim Acta 187:1–9. https://doi.org/10.1007/s00604-020-04499-x
33. Shanmugam N, Rajkamal P, Cholan S, Kannadasan N, Sathishkumar K, Viruthagiri G, Sundaramanickam A (2013) Biosynthesis of silver nanoparticles from the marine seaweed Sargassum wightii and their antibacterial activity against some human pathogens. Appl Nanosci 4(7):881–888. https://doi.org/10.1007/s13204-013-0271-4
34. Rajeshkumar S, Malarkodi C, Gnanajobitha G, Paulkumar K, Vanaja M, Kannan C, Annadurai G (2013) Seaweed-mediated synthesis of gold nanoparticles using Turbinaria conoides and its characterization. J Nanostruct Chemistry 3(1):44. https://doi.org/10.1186/2193-8865-3-44
35. Mukherjee A, Sarkar D, Sasmal S (2021) A review of green synthesis of metal nanoparticles using algae. Front Microbiol 12:693899. https://doi.org/10.3389/fmicb.2021.693899

Chapter 10
Field Deployable Pesticide Sensors: Innovations and Challenges

B. C. Indumukhi, S. Ishwarya, H. M. Deepa, H. C. Ananda Murthy, and Natthaphon Raengthon

Abstract In order to prevent pests, pesticides have been employed extensively in both industrial and agricultural operations. Thus, managing food safety greatly depends on the ability to detect pesticides with a high degree of selectivity and sensitivity. The limitations of conventional detection techniques are currently addressed by variety of field deployable pesticide sensors, like: electrochemical sensors (ES), biosensors, optical sensors, and microfluidic sensors, which have enormous promise for affordable, effective, and low detection limits using cleverly compacted devices. Public health evaluations are made much more difficult by the fact that most diseases have multiple causes and that nobody in the human population is totally immune to pesticide exposure. To lessen the effects of pesticides, it is therefore desirable to create eco-friendly pesticide substitutes and Integrated Pest Management strategies. Therefore, the purpose of this work was to provide an extensive overview of pesticides, including their forms, distribution in the environment, exposure pathways, and effects on health.

Keywords Pesticide sensors · Electrochemical sensors · Microfluidic sensors

B. C. Indumukhi (✉) · S. Ishwarya · H. M. Deepa
Department of Studies in Chemistry, Shivagangotri-Davangere University, Davangere, India

H. C. Ananda Murthy
School of Applied Sciences, Papua New Guinea University of Technology, Lae, Morobe Province, Papua New Guinea

N. Raengthon
Faculty of Science, Department of Materials Science, Chulalongkorn University, Bangkok, Thailand

R. Naik et al. (eds.), *Biocompatible Nanomaterials for Smart Sensors*, Nanostructure Science and Technology,
https://doi.org/10.1007/978-3-032-01745-1_10

10.1 Introduction

From the past few decades most of the forest parts are converted into crop lands, also modern civilization and industrialization leads to deforestation, so that soil loses its fertility. To enhance the fertility of soil, farmers utilize NPK (nitrogen, phosphorus, and potassium) fertilizers, which helps to maintain chemical, biological and physical properties of the soil. The physical state of a soil includes absorption capacity and porosity, biological state includes the presence of living microorganism, whereas chemical state includes the presence of essential nutrients and pH of the soil [1]. Fertilizers are mainly classified into two types: organic and inorganic fertilizers. Organic fertilizers are made up of plants, animal-manure, peat-moss and compost. They produce lower micro-organisms in the soil and slowdown the crop yield. Whereas inorganic fertilizers are made up of NPK, magnesium, calcium, sulfur, micronutrients (like: zinc, iron, copper, boron and manganese), etc., that enhance the crop yield. Most of the fertilizer industries are the sources for heavy metals (like Pb, As, Hg, Ni, Cr and Cd) and radionuclides (like ^{210}Po, ^{238}U and ^{232}Th). The excess consumption of nitrogen fertilizers contains nitrosamine, which may considered as carcinogenic [2].

India became the world's most populated and agriculture based country; there is need to meet the increased food demand. To fulfil the requirement needed by the people, farmers have to grow more and more crops [3]. But each year, plant pests and pathogens contribute to a significant loss of crop yields. To combat these pre-harvest losses, farmers are encouraged to extensively use fertilizers, pesticides and fungicides to enhance large scale production of crops, but these substances pose potential risks to the environment and human health [4]. Generally farmers are using chemical -based pesticides to control diseases and pests of plants thereby increasing the production of their crop yield. Based on their purpose pesticides, are classified as insecticides, herbicides, and fungicides containing different varieties of chemical composition. Most of them contains organophosphates, nitrophenol, carbamates, arsenic, and pyrethroids [5]. Continuous expose to organophosphates (OP), N-methyl carbamate, neonicotinoid, and pyrethroid pesticides specifically target nerve cells, including memory impairment, raising concerns about their rising residue levels in food, water, soil and air, which pose significant health risks. Due to their easy accessibility, organophosphates are frequently used as agricultural poisons and even in warfare [6].

Pesticides are essential in agriculture, helping farmers to boost crop yields and safeguarding them from pest-related damage. However, it is crucial to implement appropriate measures to minimize their potential environmental impact. Since their primary purpose is to target and eliminate specified organisms, continuous monitoring and evaluation of their effects on plant and animal life are necessary to ensure their responsible use while preserving ecological balance [7]. Among most of prevalent problems that afflict modern agricultural protection and regulation are pesticides, toxins, and heavy metals. Globally, heavy-metal contamination has increased and constitute significant risks to customer's health [8].

Pesticides are the toxic chemical substances used in agricultural activities to kill or repel pests, weeds, insects, and diseases from the crop. Herbicides are used to inhibit the growth of certain plant varieties, fungicides are used to destroy fungus, bactericides are used to kill bacteria, insecticides are utilized to repel insects, and rodenticides are used to kill rodents. These are some of the commonly used pesticides. The activity of pathogens, pests, and weeds lead to loss in crop yields [9–11]. Pesticides were synthesized in large scale industries, preferably petroleum byproduct are the major components of organic-based chemical pesticides. For example, DDT (dichloro-diphenyl-trichloroethane), DDE (dichloro-diphenyl-dichloroethylene), DDD (dichloro-diphenyl-dichloroethane), organochlorine were found in animal tissues, sediment, soil, and water [12]. Simultaneously, agricultural residues open-out in the surroundings causing considerable damage to the ecological communities, which toxifies soil, water, and food materials. It also affects aquatic life, food sources, aquafarming, and fisheries [13].

Continuous direct or indirect exposure to pesticides leads to health disorder in human beings as well as animals. People working in agricultural land or working in large scale pesticide production industries may come in contact with pesticides through either oral or dermal or inhalation process. As a result, they may frequently suffer from abnormalities in nervous system, nausea, dizziness, vomiting, headache, muscle weakness, and sometimes it may lead to paralysis. Pesticides containing arsenic are carcinogenic [14]. More toxic pesticides may cause serious health disorders like Hodgkin's disease, leukemia, Burkitt lymphoma, Parkinson's diseases, endocrine disruptor, etc. [15].

Hence, it is necessary to assess the pesticides present in the food products, soil, water and air even in minute quantities. Analysis of pesticides present in collected samples using conventional methods like Enzyme-linked Immunosorbent Assay (ELISA), Gas Chromatography Mass Spectrometry (GC-MS), and Highperformance liquid chromatography (HPLC), even though other chromatographic and spectrophotometric methods are reliable and sensitive, have few drawbacks under certain conditions like: sample solution preparation method is a lengthy process in GC, and some pesticides undergo decomposition in HPLC and require well-trained manpower [16].

The progress in new sensing technology becomes an alternative solution for detection of pesticides in minute quantity instead of using traditional methods. This sensing technology involves simple and easy operating techniques, cost effective, highly sensitive in monitoring the analyte sample. Now a days vigorous development has been taking place in the research area in fabricating different varieties of materials used in sensor technology with enhanced sensitivity and selectivity [17].

10.2 Types of Pesticides Based on Target Pest

Figure 10.1 shows the types of pesticides in which insecticides are type of pesticides used to kill or to control the growth of insects. For example Acephate, Actara, and Confidor are used to control bollworms, sucking pests, and hoppers. Fungicides are

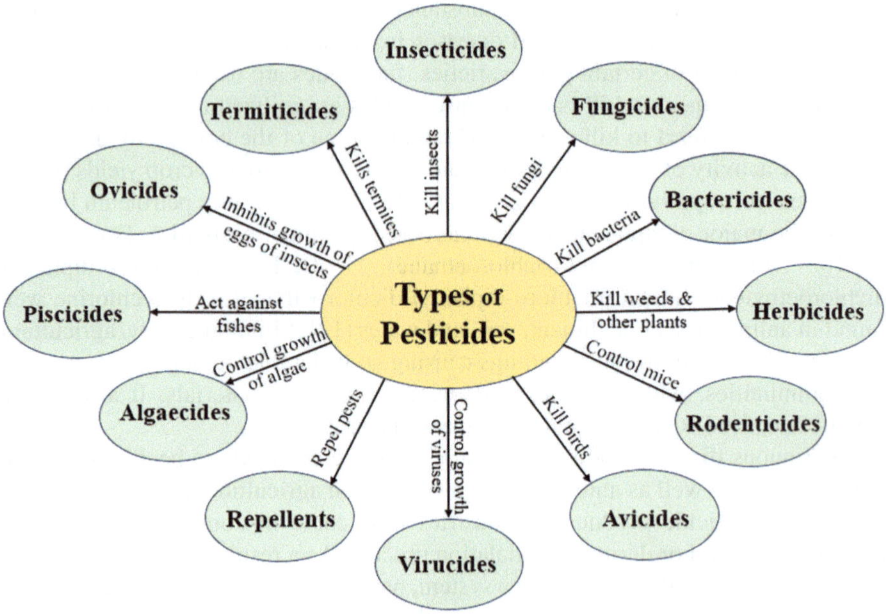

Fig. 10.1 Types of pesticides

used to prevent the growth of fungi on crop yield. Amistar, bavistin, and blitox are employed to prevent powdery-mildew (PM) and downy-mildew (DM) in vegetables and grapes, blasts, blight, rots, and wilts. Bactericides are used to kill bacteria. Benomyl, Calixin, and Folicur are some of the examples of bactericides. Gramoxone, Oxy gold, and Round-up are used as herbicides. Bromethalin, Cholecalciferol, and Zinc phosphide are used to kill rodents. 4-aminopyridine, anthraquinone, chloralose are used to repel the birds and some other pesticides are used for different purpose [3, 18–24].

10.3 Types of Field-Deployable Pesticide Sensors

10.3.1 Electrochemical Sensors (ES)

Electrochemical sensors are generally considered as analytical devises, wherein chemical reactions are transformed into quantifiable electrical impulses by electrochemical sensors. By means of redox processes, electrons transfer across the electrode surface to produce a charge in electrical characteristics, such as impedance, voltage, or current, which may be measured [25]. These sensors were employed in industries, food-safety, environment monitoring units as well as in medical devices/ diagnostic kits. Electrochemical measurements employ reference electrodes to maintain stable potential, whereas working electrodes provides precise electrochemical-measurements, hence these are used for different applications. Most commonly used reference electrodes in electrochemical measurements are silver-silver chloride (Ag/AgCl) or mercury/mercuric chloride (Hg/HgO) electrodes. Based on the type of electrochemical reactions, ES are mainly categorized into potentiometric sensors, amperometric sensors, voltametric sensors, and conductometric sensors [26, 27]. Various factors influence electrochemical reactions including structure of the material, content, size, and electrical sensitivity. Now a days graphene and its derivatives, like graphene-oxide and reduced graphene-oxide, are employed in the production of sensors [28]. Guodong Liu et al. fabricated an electrochemical immunosensor, based on nanoparticles, that has been designed to identify phosphorylated-acetylcholinesterase, a pesticide composition of organophosphate (OP) employed as insecticide as well as in chemical warfare. Organophosphate (OP) has been detected by zirconia-nanoparticles (ZrO_2-NPs) that act as specific adsorbents and the immunorecognition activities were tracked by describing monoclonal anti-phosphorylated-acetylcholinesterase antibody with ZnS@CdS quantum dots. The variables that control the electrochemical response of immun osensors, such as the concentration of QD-anti-AChE and the quantity of ZrO2-NP, were tuned. The immunosensor's voltametric response is quite linear between 10 pm and 4 nm modified AChE, with an estimated 8.0 pm limit of detection. Additionally, the immunosensor was able to identify phosphorylated AChE within human serum [29]. Congyi Zhao synthesized 3D α-Fe_2O_3-nanoflowers fabricated on gold-nanoparticles (AuNPs) along with carbon-nanotubes (CNTs) to construct sensitive, quantitative, and field-deployable biosensors for tracking organophosphate (OP) nerve toxins and pesticides [29].

10.3.2 Conductometric Sensors (CS)

Conductometric sensors (CS) track variations in a solution's electrical conductivity in relation to the concentration of analytes. In water quality studies, such as tracking the level of salts, heavy-metals, and contaminants, these sensors are helpful for detecting ionic species. Since conductometric sensors do not need a reference electrode, their design is simpler and less expensive than that of amperometric and potentiometric sensors. Applications of biosensing use conductometric sensors to identify biomolecules like proteins and DNA by measuring how a solution's conductivity changes in response to molecular interactions. Recent developments in nanostructuring materials, such as conductometric sensors based on graphene, have increased their specificity and sensitivity for various uses.

These sensors rely on changes in a thin layers or a vast amount of material's electrical conductivity, which are influenced by the material under study. Conductimetric methods lack fundamental selectivity. Thanks to significantly improved instrumentation and changes to active surfaces for selectivity, these sensor design strategies are now more practical. Since conductimetric methods do not require reference electrodes, they are simple and affordable, making them appealing for real-world applications. The measurement of conductivity has been the exclusive reliant on improved instruments. We will start by looking at the most common materials used in these sensors. The methods for numerically processing the analytical signal will then be looked at, as well as specific events or traits that are appropriate for sensing [30].

10.3.3 Amperometric Sensor

The applied voltage acts as an accelerator for electrocatalytic redox processes, which produces electric currents proportional to the analyte concentration. Amperometric observations are widely employed as an analytical technique with excellent precision as well as sensitiveness. The basic instrumentation demands a controlled-potential arrangement. Two electrodes immersed in an electrolyte of the proper composition make up an electrochemical cell. Using a three-electrode cell, in which one electrode serves as a reference electrode, like silver/silver chloride electrode (Ag/AgCl) or mercury/mercurous chloride (Hg/Hg_2Cl_2) electrode, that maintains constant-potential [31].

Another electrode functions as a working electrode that identifies an analyte molecule. A static conducting material, such as graphite or platinum, is frequently used as an auxiliary electrode. A supporting electrolyte maintains the constant ionic strength in controlled-potential investigations by preventing electromigration effects. Accuracy and sensitivity have been thoroughly discussed in theoretical and practical ways, where the applied voltage acts as a catalyst for electrocatalytic redox reactions that produce electrical currents proportionate to the analyte concentration [32]. The working electrode, also known as the sensor, is maintained at a steady voltage, while the current is being tracked throughout an amperometric assessment. The concentration of the analyte present is then linked to the current [26].

10.3.4 Voltametric Sensors

Voltametric sensors are an effective technique employed for examining the process of analysis of interface events that take place on the electrode surface is cyclic voltammetry (CV), which uses a potential scan. The shift in the current peaks can be used to determine the redox process of electrically active species. This allows for the quantitative determination of pesticide concentrations [33]. The basis for voltammetric sensor functions is the measurement of the current responses that depends on the applied voltage being sweep. This offers crucial information about the analytes' electrochemical reaction [34]. The redox responses at the surface of an electrode are investigated using a variety of voltametric techniques, including square wave, differential pulse, and cyclic voltammetry. An electrochemical signal is applied across the working electrode as part of the voltammetry working principle. Additionally, the measured current is recorded. Distinct peaks in the measured current are caused by the oxidation or reduction of the targeted analyte at certain potential values due to variations in the scanned values of the applied potential [35, 36].

10.3.5 Biosensors

Biosensors are composed of bicomponent analytic probes. One of the component comprises biological elements like cells, DNA, enzymes, microorganism, aptamers, receptors, and some of the selective antibodies. Another component—a transducer—converts biological information (analyte) into physical signals. Biosensors provide real-time response to an analyte, hence they are employed in toxicity assessment. Electrochemical, optical, thermometric, and piezoelectric are few examples for the transducers used in biosensors [37]. Biosensors were designed for cost effective, miniaturization, in situ finding, and quick detection with highest sensitivity. Decreasing the use of biosensors is due to their microfluidic environment, which diminishes the scattering distance within the biological element and the analyte. It will help for quick assessment of the analyte molecule. During some of the interactions heat will be released, which can be measured quantitatively. In few interactions color change may occur or redox reaction may occur on metal surfaces [7].

10.3.6 Optical Sensors: Colorimetric Sensors

In the visible portion of the electromagnetic-spectrum, colloidal metal-nanoparticles with an elevated extinction-coefficient show distinct colors. The analytical characteristics of metal NPs are strongly reliant on the size and form. When colloidal nanoparticles interact with target analytes, their color changes, which is the fundamental idea behind calorimetric techniques [38]. The major development in colorimetric technology based on nanoparticles over conventional methods are quick, easy, and sensitive [39, 40]. After mixing the small number of nanoparticles with the analytes, the presence of the analytes is confirmed in a matter of minutes by a color shift. Analyte identification uses a wide variety of nanomaterials. Noble metal-nanoparticles are a stronger candidate for colorimetric analyte identification due to their superior extinction coefficient and surface plasmon resonance [41, 42]. The development of enhanced sensing systems and portable instruments that integrate nanotechnology-enabled solutions was made possible by improvements in the controlled-size, apparent characteristics, and composition of nanoparticle structures [43]. Numerous nanostructures, including graphene, carbon nanotubes, palladium, gold, silica, silver, cerium oxide, and magnetic-nanoparticles, are employed in sensing techniques based on nanotechnology [44, 45]. Special chemical and optical characteristics, including self-assembly, localized surface-plasmon resonance (LSPR), bio-compatibility, and resonance light-scattering (RLS), are attributed to these nanostructures, respectively [46].

Fang Li et al. synthesized platinum-nanoparticles (Pt NPs) as a nanozyme sensory receptor; a straightforward wideband colorimetric sensor array was built for pesticide recognition and detection. Pt-NPs directly oxidized 3,3′,5,5′-tetramethylbenzidine (TMB), three significantly increased adsorption peaks were produced at 370, 450, and 650 nm, which provides three reactive sensing elements for detection [47]. Dejun Xiong et al. synthesized very stable silver-nanoparticles altered with p-sulfonatocalix[n]arene (where $n = 4, 8$) using one-pot method in aqueous environments. Compared to p-sulfonatocalix[8]arene altered silver-nanoparticles, p-sulfonatocalix[4]arene altered silver-nanoparticles can be used as a novel colorimetric probe for optunal, enabling a quick quantitative-assay of optunal lower to a concentration of 10^{-7} M. This indicates a great deal of potential for use in real-time in situ optunal detection [48].

10.3.7 Fluorescence Sensor

Freely moving electrons between the ground state and the excited state can be promoted by absorbing enough energy. Conversely, this transfer can be carried out starting at the lowest vibration level. The Stokes' shift process is introduced as an excited state following a nonradiative-transition of internal transformation or relaxation of vibrations. Through a variety of signal transduction mechanisms, such as distinctive wavelength (blue/red-shift), lifetime, fluorescence-brightness (quenching/enhancement), and anisotropy, the selective interactions can affect the output particular fluorescence and fluorescence unique information of sensors [49, 50].

In analytical signal transduction, fluorescence probes function as participants or markers of physical/chemical transitions in states. The stability, sensibility, and ability of established procedures to prevent interference are all greatly impacted by the physico-chemical characteristics of these compounds. For example, oxygen may readily oxidize nanoclusters created by the reduction of silver (Ag) and copper (Cu), which eventually causes them to lose their luminous characteristics. The characteristics of various materials decide which molecules can either impact the experiment's stability (e.g., solubility influences the system's dispersion, ligands with hydrophobic properties causes tiny particles surfaces tend to agglomerate in water-based solution), have an enhancement impact [51]. Hanqing Wu et al. reported a fluorescent sensor for the identification of starane pesticide (thiourea-bridged-bis-tetraphenylimidazole). It showed good potential applicability for identification of starane in everyday life and agriculture [52].

10.3.8 *Microfluidic Sensors (MS) or "Lab-on-a-Chip"*

Microfluidic sensors are those that control the rate of flow fluids at micro-scale levels. They show high sensitivity, rapid response, and are inexpensive, hence employed in environmental assessment and healthcare instruments. They have capability to assess temperature, flow rate, and pH of the fluids [53]. The major component of MS are holes, carries fluids through microchannels to execute reaction. These sensors are integrated with driving-units, control unit, detector, and amplifier to detect and amplify the signals. In general electrochemical and optical methods were employed for detection of an analyte. Usually optical methods like colorimetric-method, chemiluminescence, or Raman spectroscopic methods were employed to monitor the pesticides, whereas electrochemical analysis method involves potentiometric, amperometric, and voltametric techniques [54, 55]. Compared to optical methods, electrochemical methods reveal high sensitivity toward pesticide detection even with miniaturizing the MS electrode. The flow of fluids in microchannels can be characterized by Reynold's (Re) number, which is dependent on distinct flow-geometries like T-shape, Y-shape, and linear and turbulent geometries. Reynold's number can be calculated by the expression:

$$R_e = \frac{\rho V D}{\mu}$$

where R_e is Reynold's number, ρ is density of a fluid, V is average-velocity of fluid, D is diameter of microchannel, and μ is viscosity of fluid. Guoying Hao et al. fabricated a quantum-dot based dual-channel microfluidic sensor, as shown in the Fig. 10.2 [56].

The development of microchannels on the surface of a paper to direct the analyte flow is a crucial stage in the creation of microfluidic sensor chip. The following techniques could be used to create the micro-channels: (1) linear cut to create grooves; and (2) creating a physical barrier by combining the permeable polymer with polyurethane, polystyrene, or polydimethylsiloxane. The disadvantage is that hydrophobic compounds like alkylketene dimer, photoresist, and paraffin on

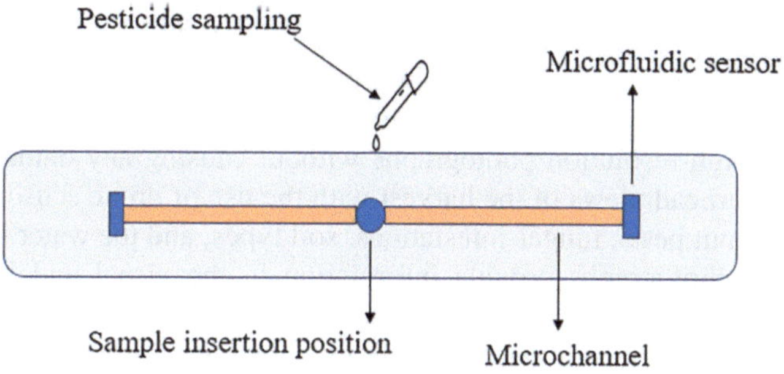

Fig. 10.2 Double-channel microfluidic pesticide sensor

filter-paper can be modified to form channels, which may prevent lipophilic molecules like cholesterol from being analyzed in the channels. This paper synthesizes the microfluidic paper chip using the third approach [57].

10.4 Innovations in Pesticide Sensor Technologies

10.4.1 Advancements in Nanotechnology for Improved Sensitivity and Selectivity

Latest advancements in nanotechnology aid in the efficient transportation of agricultural chemicals/pesticides, resistance against unfavorable weather, regulated discharge of active-ingredients, potential enhancement by lowering the need for active-ingredients, and better storage. These elements increase the biopesticides' effectiveness and durability. For instance, Bt-crystals and spores were reduced to nanoscale in size. It was further prepared as a suspension-concentrate after being reduced in size by grinding. As a result, Bt's effectiveness over *Spodpotera litura* on castor was enhanced by 2.18 to 2.24 times, matching that of the synthetic insecticide profenophos [58].

The regulated release of insecticides into the soil is facilitated by nanopesticides, which lowers the need for insecticides to control the bug. Compared to standard compositions, the imidacloprid-nanopesticide based on polyethylene-glycol aids in a slower release of insecticide/pesticide. Likewise, a number of insecticides, including beta-cyfluthrin, carbofuran, emamectin-benzoate, thiram, and thiamethoxam, were coated with polymers for photostability and delayed release [59].

Drone mapping is regarded as the most crucial technology for accuracy in agriculture, pesticide spraying, crop health surveillance, and innovative farming. Using drones for aerial mapping makes use of photos taken at various wavelengths and calculates indices of vegetation to identify various crop situations [60]. In previous decades, satellites or manned airplanes were employed to obtain the necessary photographs for precision farming [61]. It is highly expensive to take pictures with manned aircraft, and the issue with satellite photography is that, in most cases, the geographic details of the photos are not as good as it could be. Furthermore, the weather has an impact on the quality and availability of photographs [62, 63]. Crop monitoring has changed as a result of improvements in drone capabilities and the decreased weight of carrying devices. This procedure saves time, is inexpensive, and produces high-resolution photographs without causing any damage. Farmers can observe overhead views of the harvest with the use of drone sensors. This provides details about pests, fungal-infestations, soil types, and the water flow. Drone-collected crop photographs provide information in the visual and infrared (IR) spectrums. It is possible to extract many characteristics from these photos, providing details about plant health that are invisible to the human eye [64, 65].

IoT usage is accelerating in a variety of industries, including healthcare, safety, and agriculture [66]. The farming industry will become more sophisticated as a result of IoT combined with robots and AI, which will also reduce the number of manual laborers from 90% to 10%. India is believed to be the site of one of the largest agricultural scientific endeavors in the world [67]. Numerous developments in contemporary technology and ingenuity have led to breakthrough growths in agriculture. The use of modern techniques in the agricultural sector, such as IoT, detectors, analysis of large quantities of data, mobile computing, and cloud services, has been studied in the past [68, 69]. The results of a study on an IoT-based process in the agricultural context were astounding [70].

10.4.2 Challenges in Deploying Pesticide Sensors in the Field

The controlled manufacture of nanoparticles, which have significant uses in all branches of research, is known as nanotechnology. The fundamental difficulties faced in nanotechnology are reproducibility, particle size and shape control, and scalability, which are all aspects of nanoparticle production. Repetitive production of nanoparticles makes it challenging to get the appropriate size and form. Metal precursors are reduced and stabilized using various agents in biological, chemical, and green-synthesis techniques to create nanoparticles [71]. Since the size and form of nanoparticles affect their optical characteristics, choosing the right synthesis technique is crucial for the creation of colorimetric sensors. For colorimetric applications, spherical and small nanoparticles are ideal [46].

The durability of nanoparticles is a major obstacle that continues to impede the quick development and efficacy of pesticide sensors based on nanoparticles. The majority of the synthesized nanoparticles utilized in colorimetric applications are unstable and can aggregate. Unfortunately, the environment contains a lot of interfering compounds that typically cause nanoparticles to aggregate. In addition to this issue, the stability of nanoparticles is impacted by the existence of sugars, suspended-solids, proteins, fats, nucleic-acids, and other substances found in food and environments, making it challenging to identify residual-pesticides from food matrix [72].

Different analytical methods that employ nanoparticles offer good selectivity as well as sensitivity for pesticide measurement; a number of factors still affect metal nanoparticle sensitivity [73]. The most researched parameter for detecting residual pesticides is the limit of detection (LOD). The kinds of matrix, surface-characteristics, and synthesis of the nanoparticles all effect on the limit of detection. The maximum-residue-limit (MRL) and the LOD for determining the presence of pesticides in parts-per-million (ppm) and parts-per-billion (ppm) levels cannot be compared. The maximum number of pesticides allowed in food samples is known as the maximum-residue-limit (MRL), and it varies according to the kind and makeup of the product [42].

10.5 Future Prospects and Research Directions

There are a number of obstacles in the creation and use of field deployable pesticide sensors like:

- Sensor durability and reproducible nature: ES have longevity problems, whereas signal drifting and corrosion of electrodes lead to the sensor's gradual degradation.
- Disturbance from non-target compounds: Identical compounds in the intricate matrix of samples cause disturbance for the majority of electrochemical sensors, which results in a selectivity issue.
- The connection with AI and machine learning: While AI has promise for enhancing sensor data analysis, further research is required before it can be fully implemented in pesticide sensing platforms.

10.6 Conclusion

Innovations for monitoring pesticides are a game-changer in agriculture, offering answers to immediate problems pertaining to sustainable development, food safety, and regulatory compliance. Biochemical monitoring has been entirely modified by a variety of field deployable pesticide sensors. They have demonstrated their usefulness in a number of industries including workplace protection, ecological observing, and healthcare. Improvements regarding computer integration, methods of production, and nanostructures have raised the degree of efficiency, selectivity, and sensitivity of these sensors. Still there are significant obstacles in the areas of interference, repeatability, and sensor stability. Further crucial elements for sensor advancement will include the creation of multi-analyte sensing devices, with AI incorporation.

References

1. Sharma A, Chetani R (2017) A review on the effect of organic and chemical fertilizers on plants. Int J Res Appl Sci Eng Technol 5(2):677–680
2. Savci S (2012) Investigation of effect of chemical fertilizers on environment. Apcbee Procedia 1:287–292. https://doi.org/10.1016/j.apcbee.2012.03.047
3. Pandey JK (2023) Use of hazardous chemical pesticides in India: a review. Proc Natl Acad Sci India Sect B Biol Sci 93(3):523–537. https://doi.org/10.1007/s40011-022-01425-4
4. Shad NA, Munawar A, Javed Y, Rakha A, Riaz A, Din SU, Salman M (2023) In-field deployable and facile nanosensor for the detection of pesticides residues. Anal Chim Acta 1259:341204. https://doi.org/10.1016/j.aca.2023.341204
5. Aragay G, Pino F, Merkoçi A (2012) Nanomaterials for sensing and destroying pesticides. Chem Rev 112(10):5317–5338

6. Padilla S (2006) Cumulative effects of organophosphorus or carbamate pesticides. In: Toxicology of organophosphate & carbamate compounds. Academic Press, New York, pp 607–615. https://doi.org/10.1016/B978-012088523-7/50042-9
7. Mutunga T, Sinanovic S, Harrison CS (2024) Integrating wireless remote sensing and sensors for monitoring pesticide pollution in surface and groundwater. Sensors 24(10):3191. https://doi.org/10.3390/s24103191
8. Dhamu VN, Poudyal D, Samson M, Paul A, Muthukumar S, Prasad S (2023) Environmental biosensors for agro-safety based on electrochemical sensing mechanism with an emphasis on pesticide screening. ECS Sensors Plus 2(2):024601. https://doi.org/10.1149/2754-2726/acde5d
9. Sharma A, Kumar V, Shahzad B, Tanveer M, Sidhu GPS, Handa N, Thukral AK (2019) Worldwide pesticide usage and its impacts on ecosystem. SN Appl Sci 1:1–16. https://doi.org/10.1007/s42452-019-1485-1
10. Oerke EC (2006) Crop losses to pests. J Agric Sci 144(1):31–43. https://doi.org/10.1017/S0021859605005708
11. Savary S, Bregaglio S, Willocquet L, Gustafson D, Mason D'Croz D, Sparks A, Garrett K (2017) Crop health and its global impacts on the components of food security. Food Secur 9:311–327. https://doi.org/10.1007/s12571-017-0659-1
12. Warne MSJ, Reichelt-Brushett A (2023) Pesticides and biocides. In: Marine pollution–monitoring, management and mitigation. Springer, Cham, pp 155–184
13. Carvalho FP (2017) Pesticides, environment, and food safety. Food Energy Secur 6(2):48–60. https://doi.org/10.1002/fes3.108
14. Ben-Zur R, Hake H, Hassoon S, Bulatov V, Schechter I (2011) Optical analytical methods for detection of pesticides. Rev Anal Chem 30(3–4):123–139. https://doi.org/10.1515/REVAC.2011.104
15. Bilal M, Iqbal HM, Barceló D (2019) Persistence of pesticides-based contaminants in the environment and their effective degradation using laccase-assisted biocatalytic systems. Sci Total Environ 695:133896. https://doi.org/10.1016/j.scitotenv.2019.133896
16. Verma N, Bhardwaj A (2015) Biosensor technology for pesticides—a review. Appl Biochem Biotechnol 175:3093–3119
17. Sabherwal P, Mutreja R, Suri CR (2016) Biofunctionalized carbon nanocomposites: new-generation diagnostic tools. TrAC Trends Anal Chem 82:12–21. https://doi.org/10.1016/j.trac.2015.10.006
18. Gupta RC, Mukherjee IRM, Malik JK, Doss RB, Dettbarn WD, Milatovic D (2019) Insecticides. In: Biomarkers in toxicology. Academic Press, New York, pp 455–475. https://doi.org/10.1016/B978-0-12-814655-2.00026-8
19. Jie HOU, Li Xi WANG, Jin Ze WANG, Li Yuan CHEN, Bing Jun HAN, Tong YANG, Wen Xin LIU (2025) Insights on common fungicides: a national survey on farmland soils from mainland China. J Hazard Mater 492:138177. https://doi.org/10.1016/j.jhazmat.2025.138177
20. Kim KH, Kabir E, Jahan SA (2017) Exposure to pesticides and the associated human health effects. Sci Total Environ 575:525–535. https://doi.org/10.1016/j.scitotenv.2016.09.009
21. Yadav IC, Devi NL (2017) Pesticides classification and its impact on human and environment. Environ Sci Eng 6(7):140–158
22. Kaur R, Mavi GK, Raghav S, Khan I (2019) Pesticides classification and its impact on environment. Int J Curr Microbiol Appl Sci 8(3):1889–1897. https://doi.org/10.20546/ijcmas.2019.803.224
23. Pest HTIY, Pesticides LR, Poisonings H, Land IOT. Topic fact sheet
24. Barakat NM, Hamdy MM, El-Gendy AA (2025) Repellent effect of captan 50% WP and bayfidan 25% EC fungicides against house sparrow bird, passer Domesticus niloticus (L.) under laboratory and field conditions. Menoufia J Plant Prot 10(1):13–19. https://doi.org/10.21608/mjapam.2025.352258.1045
25. Tajik S, Dourandish Z, Nejad FG, Beitollahi H, Jahani PM, Di Bartolomeo A (2022) Transition metal dichalcogenides: synthesis and use in the development of electrochemical sensors and biosensors. Biosens Bioelectron 216:114674. https://doi.org/10.1016/j.bios.2022.114674

26. Kimmel DW, LeBlanc G, Meschievitz ME, Cliffel DE (2012) Electrochemical sensors and biosensors. Anal Chem 84(2):685–707. https://doi.org/10.1021/ac202878q
27. Dincer C, Bruch R, Costa-Rama E, Fernández-Abedul MT, Merkoçi A, Manz A, Güder F (2019) Disposable sensors in diagnostics, food, and environmental monitoring. Adv Mater 31(30):1806739. https://doi.org/10.1002/adma.201806739
28. Tanwar S, Mathur D (2021) Graphene-based nanocomposites as sensing elements for the electrochemical detection of pesticides: a review. J Solid State Electrochem 25(8):2145–2159. https://doi.org/10.1007/s10008-021-04990-2
29. Liu G, Wang J, Barry R, Petersen C, Timchalk C, Gassman PL, Lin Y (2008) Nanoparticle-based electrochemical immunosensor for the detection of phosphorylated acetylcholinesterase: an exposure biomarker of organophosphate pesticides and nerve agents. Chemistry 14(32):9951–9959. https://doi.org/10.1002/chem.200800412
30. Chen W, Cai S, Ren QQ, Wen W, Zhao YD (2012) Recent advances in electrochemical sensing for hydrogen peroxide: a review. Analyst 137(1):49–58. https://doi.org/10.1039/C1AN15738H31
31. Hu J, Sang G, Zeng N, Lv C, Xu C (2022) Amperometric sensor for the detection of hydrogen stable isotopes based on Pt nanoparticles confined within single-walled carbon nanotubes (SWNTs). Sens Actuators B Chem 356:131344. https://doi.org/10.1016/j.snb.2021.131344
32. Stasyuk NY, Gayda GZ, Zakalskiy AE, Fayura LR, Zakalska OM, Sibirny AA, Gonchar MV (2022) Amperometric biosensors for L-arginine and creatinine assay based on recombinant deiminases and ammonium-sensitive cu/Zn (hg) S nanoparticles. Talanta 238:122996. https://doi.org/10.1016/j.talanta.2021.122996
33. Sun X, Du S, Wang X (2012) Amperometric immunosensor for carbofuran detection based on gold nanoparticles and PB-MWCNTs-CTS composite film. Eur Food Res Technol 235(3):469–477. https://doi.org/10.1007/s00217-012-1774-z
34. Power AC, Gorey B, Chandra S, Chapman J (2018) Carbon nanomaterials and their application to electrochemical sensors: a review. Nanotechnol Rev 7(1):19–41. https://doi.org/10.1515/ntrev-2017-0160
35. Simões FR, Xavier MG (2017) Electrochemical sensors. Nanosci Appl 1099:155–178
36. Dwivedi CP, Kumar D (2025) Electrochemical sensors: a review of developments and applications. Indian J Pharm Chem Anal Tech 1(1):13–27
37. Van Emon JM, Drum T, Dill K (2019) Immunoassays and biosensors. In: Analysis of pesticides in food and environmental samples, 2nd edn. CRC Press, Boca Raton, pp 105–134
38. Ullah N, Mansha M, Khan I, Qurashi A (2018) Nanomaterial-based optical chemical sensors for the detection of heavy metals in water: recent advances and challenges. TrAC Trends Anal Chem 100:155–166. https://doi.org/10.1016/j.trac.2018.01.002
39. Singh R, Thakur P, Thakur A, Kumar H, Chawla P, Rohit V, Kumar N (2021) Colorimetric sensing approaches of surface-modified gold and silver nanoparticles for detection of residual pesticides: a review. Int J Environ Anal Chem 101(15):3006–3022. https://doi.org/10.1080/03067319.2020.1715382
40. Schoolaert E, Hoogenboom R, De Clerck K (2017) Colorimetric nanofibers as optical sensors. Adv Funct Mater 27(38):1702646. https://doi.org/10.1002/adfm.201702646
41. Zhang D, Liu Q (2016) Biosensors and bioelectronics on smartphone for portable biochemical detection. Biosens Bioelectron 75:273–284. https://doi.org/10.1016/j.bios.2015.08.037
42. Singh R, Kumar N, Mehra R, Kumar H, Singh VP (2020) Progress and challenges in the detection of residual pesticides using nanotechnology based colorimetric techniques. Trends Environ Anal Chem 26:e00086. https://doi.org/10.1016/j.teac.2020.e00086
43. Giannoulis KM, Giokas DL, Tsogas GZ, Vlessidis AG (2014) Ligand-free gold nanoparticles as colorimetric probes for the non-destructive determination of total dithiocarbamate pesticides after solid phase extraction. Talanta 119:276–283. https://doi.org/10.1016/j.talanta.2013.10.063

44. Rastogi L, Dash K, Ballal A (2017) Selective colorimetric/visual detection of Al3+ in ground water using ascorbic acid capped gold nanoparticles. Sensors Actuators B Chem 248:124–132. https://doi.org/10.1016/j.snb.2017.03.138
45. Kumar N, Kumar H, Mann B, Seth R (2016) Colorimetric determination of melamine in milk using unmodified silver nanoparticles. Spectrochim Acta A Mol Biomol Spectrosc 156:89–97. https://doi.org/10.1016/j.saa.2015.11.028
46. Chawla P, Kaushik R, Swaraj VS, Kumar N (2018) Organophosphorus pesticides residues in food and their colorimetric detection. Environ Nanotechnol Monit Manage 10:292–307. https://doi.org/10.1016/j.enmm.2018.07.013
47. Li F, Jiang J, Peng H, Li C, Li B, He J (2022) Platinum nanozyme catalyzed multichannel colorimetric sensor array for identification and detection of pesticides. Sensors Actuators B Chem 369:132334. https://doi.org/10.1016/j.snb.2022.132334
48. Xiong D, Li H (2008) Colorimetric detection of pesticides based on calixarene modified silver nanoparticles inwater. Nanotechnology 19(46):465502. https://doi.org/10.1088/0957-4484/19/46/465502
49. Yang Q, Li J, Wang X, Peng H, Xiong H, Chen L (2018) Strategies of molecular imprinting-based fluorescence sensors for chemical and biological analysis. Biosens Bioelectron 112:54–71. https://doi.org/10.1016/j.bios.2018.04.028
50. Wu Y, Sun J, Huang X, Lai W, Xiong Y (2021) Ensuring food safety using fluorescent nanoparticles-based immunochromatographic test strips. Trends Food Sci Technol 118:658–678. https://doi.org/10.1016/j.tifs.2021.10.025
51. Ren H, Mei Tai S, Barimah AO, Mao M, Peng C, Xu J, Wang Z (2024) Advancement of pesticides fluorescence detection: from sensing strategies to application prospect. Trends Food Sci Technol 152:104682. https://doi.org/10.1016/j.tifs.2024.104682
52. Wu H, Chen H, Zheng S, Guo H, Yang F (2022) First fluorescence sensor for detecting pesticide starane. Sensors Actuators B Chem 369:132347. https://doi.org/10.1016/j.snb.2022.132347
53. Weng X, Neethirajan S (2017) Ensuring food safety: quality monitoring using microfluidics. Trends Food Sci Technol 65:10–22. https://doi.org/10.1016/j.tifs.2017.04.015
54. Yao J, Liu Z, Jin M, Zou Y, Chen J, Xie P, Shui L (2020) Uniform honeycomb CNT-microparticles prepared via droplet-microfluidics and sacrificial nanoparticles for electrochemical determination of methyl parathion. Sensors Actuators B Chem 321:128517. https://doi.org/10.1016/j.snb.2020.128517
55. Kaur H, Kumari N, Sharma A, Sachdeva M, Mutreja V (2022) Optical and electrochemical microfluidic sensors for water contaminants: a short review. Mater Today Proc 48:1673–1679. https://doi.org/10.1016/j.matpr.2021.09.534
56. AlMashrea BA, Almehdi AM, Damiati S (2024) Simple microfluidic devices for in situ detection of water contamination: a state-of-art review. Front Bioeng Biotechnol 12:1355768. https://doi.org/10.3389/fbioe.2024.1355768
57. Hao G, Zhang Z, Ma X, Zhang R, Qin X, Sun H, Rong J (2020) A versatile microfluidic paper chip platform based on MIPs for rapid ratiometric sensing of dual fluorescence signals. Microchem J 157:105050. https://doi.org/10.1016/j.microc.2020.105050
58. Vineela V, Nataraj T, Reddy G, Vimala Devi PS (2017) Enhanced bioefficacy of bacillus thuringiensis var. kurstaki against Spodoptera litura (Lepidoptera: Noctuidae) through particle size reduction and formulation as a suspension concentrate. Biocontrol Sci Tech 27(1):58–69. https://doi.org/10.1080/09583157.2016.1247433
59. Sujayanand GK, Sheelamary S, Prabhu G (2021) Recent innovations and approaches for insect pest management in agriculture. Biotica Res Today 3(2):100–102
60. Akram T, Naqvi SR, Haider SA, Kamran M (2017) Towards real-time crops surveillance for disease classification: exploiting parallelism in computer vision. Comp Electr Eng 59:15–26. https://doi.org/10.1016/j.compeleceng.2017.02.020
61. Liaghat S, Balasundram SK (2010) A review: the role of remote sensing in precision agriculture. Am J Agric Biol Sci 5(1):50–55

62. Hunt ER Jr, Daughtry CS (2018) What good are unmanned aircraft systems for agricultural remote sensing and precision agriculture? Int J Remote Sens 39(15–16):5345–5376. https://doi.org/10.1080/01431161.2017.1410300
63. Murugan D, Garg A, Singh D (2017) Development of an adaptive approach for precision agriculture monitoring with drone and satellite data. IEEE J Select Top Appl Earth Observ Remote Sens 10(12):5322–5328. https://doi.org/10.1109/JSTARS.2017.2746185
64. Puri V, Nayyar A, Raja L (2017) Agriculture drones: a modern breakthrough in precision agriculture. J Stat Manag Syst 20(4):507–518. https://doi.org/10.1080/09720510.2017.1395171
65. El Bilali H, Allahyari MS (2018) Transition towards sustainability in agriculture and food systems: role of information and communication technologies. Inf Process Agric 5(4):456–464. https://doi.org/10.1016/j.inpa.2018.06.006
66. Kour VP, Arora S (2020) Recent developments of the internet of things in agriculture: a survey. IEEE Access 8:129924–129957. https://doi.org/10.1109/ACCESS.2020.3009298
67. Wani JA, Sharma S, Muzamil M, Ahmed S, Sharma S, Singh S (2022) Machine learning and deep learning based computational techniques in automatic agricultural diseases detection: methodologies, applications, and challenges. Arch Comput Methods Eng 29(1):641–677. https://doi.org/10.1007/s11831-021-09588-5
68. Delnevo G, Girau R, Ceccarini C, Prandi C (2021) A deep learning and social iot approach for plants disease prediction toward a sustainable agriculture. IEEE Internet Things J 9(10):7243–7250. https://doi.org/10.1109/JIOT.2021.3097379
69. Wang Y, Wang H, Peng Z (2022) Rice diseases detection and classification using attention based neural network and bayesian optimization. arXiv preprint arXiv:2201.00893. https://doi.org/10.48550/arXiv.2201.00893
70. Mukhopadhyay SC, Tyagi SKS, Suryadevara NK, Piuri V, Scotti F, Zeadally S (2021) Artificial intelligence-based sensors for next generation IoT applications: a review. IEEE Sensors J 21(22):24920–24932. https://doi.org/10.1109/JSEN.2021.3055618
71. Ealia SAM, Saravanakumar MP (2017, November) A review on the classification, characterisation, synthesis of nanoparticles and their application. IOP Conf Ser Mater Sci Eng 263(3):032019. https://doi.org/10.1088/1757-899X/263/3/032019
72. Campos-Mañas MC, Plaza-Bolaños P, Martínez-Piernas AB, Sánchez-Pérez JA, Agüera A (2019) Determination of pesticide levels in wastewater from an agro-food industry: target, suspect and transformation product analysis. Chemosphere 232:152–163. https://doi.org/10.1016/j.chemosphere.2019.05.147
73. Wen D, Liu W, Herrmann AK, Haubold D, Holzschuh M, Simon F, Eychmüller A (2016) Simple and sensitive colorimetric detection of dopamine based on assembly of cyclodextrin-modified au nanoparticles. Small 12(18):2439–2442. https://doi.org/10.1002/smll.201503874/abstract

Chapter 11
Nanostructured Electrochemical Sensors for Antibiotic Detection

S. Ishwarya, B. C. Indumukhi, H. M. Deepa, H P Nagaswarupa, Ramachandra Naik, Kar Ban Tan, K G Basava Kumar, and Janitha Abeywickrema Liyanage

Abstract The aquatic environment is known to be seriously contaminated by antibiotic residues on a worldwide basis. In the world, antibiotics are widely used as a major class of pharmaceutically active chemicals (PhACs). Manures from animal husbandry, hospitals, municipal sewage, and home disposal are the main sources of these wastes. Surface and drinking waters, sewage effluents, soils, sediments, and a variety of plant species in nations including China, Japan, South Korea, Europe, the United States, Canada, and India are all often found to contain these residues. A significant amount of antibiotics are released into the environment as metabolites in urine and feces, despite their usage as medications in both people and animals. As sensitive and quantitative analytical methods have improved, antibiotics at quantities ranging from nanograms per liter (ng/L) to milligrams per liter (mg/L) have been regularly detected in environmental matrices. Because agricultural soils have such high particle adsorption capabilities, they are a major reservoir for antibiotic residues. There are major worries about the potential for human and animal intake of these medications due to the fact that plants cultivated in soils irrigated with water tainted with PhAC can absorb and deposit them in a variety of tissues, including roots, leaves, and fruits. This research concludes with an assessment and synopsis of commercially available and spin-off antibiotic detection sensors, highlighting

S. Ishwarya (✉) · B. C. Indumukhi · H. M. Deepa · H P Nagaswarupa
Department of Studies in Chemistry, Davangere University, Davangere, Karnataka, India

R. Naik
Department of Physics, New Horizon College of Engineering, Bangalore, India

K. B. Tan
Faculty of Science, Department of Chemistry, Universiti Putra Malaysia, Serdang, Malaysia

K G Basava Kumar
Department of R&D, IPR Cell, ATAL-IDEA LAB, Global Campus, VTU, Belagavi, India

J. A. Liyanage
Department of Chemistry, University of Kelaniya, Colombo, Sri Lanka

R. Naik et al. (eds.), *Biocompatible Nanomaterials for Smart Sensors*, Nanostructure Science and Technology,
https://doi.org/10.1007/978-3-032-01745-1_11

their portability and versatility, through elucidating the state, function, and prospects of electrochemical biosensors in antibiotic detection.

Keywords Antibiotics · Sensors · Nanomaterials · Biosensor

11.1 Introduction

One major concern associated with food safety is related to residual effects of antibiotics that are widely used to treat animals and result in antimicrobial resistance. Antibiotics are a widely used class of antimicrobial agents employed in both human and veterinary medicine to treat various infectious diseases [1, 2]. Beyond medical applications, antibiotics play a crucial role in animal husbandry, where they are used not only for treating illnesses but also for promoting growth, enhancing feed efficiency, and preventing infections [3–5]. Initially introduced to combat and prevent diseases in livestock, veterinary antibiotics have since been incorporated into animal feed to regulate reproductive cycles, improve breeding performance, and serve as prophylactic agents and growth promoters—significantly surpassing their use for therapeutic purposes [2, 6]. The selection and consumption of antibiotics vary globally, influenced by factors such as the type of livestock, farming intensity, production objectives, regional agricultural practices, and the presence (or lack) of clear regulatory policies governing antibiotic usage [7, 8]. In many developing and underdeveloped nations, the use of antibiotics as growth promoters in animal feed remains largely unregulated [9]. A major concern is that the antibiotics used in veterinary and agricultural settings often belong to the same classes as those prescribed for human use, functioning in a similar manner and posing risks of cross-resistance [10]. On a global scale, the estimated annual use of antimicrobial agents per kilogram of animal production varies by species, with cattle requiring approximately 45 mg/kg [7]. The administration methods of these antibiotics differ based on the animal species and farming practices. The expansion of large-scale farming operations, inadequate government regulations, poor enforcement of antibiotic control policies, ineffective infection prevention measures, and resistance from farmers to adopting alternative farming methods all contribute to the continued use of unnecessary antibiotics in livestock production [11]. Antibiotics are still commonly used as growth enhancers to protect animal health, boost productivity, and increase farm profitability [12].

In developing nations, many farmers purchase antibiotics over the counter and engage in multidrug practices, significantly influencing the volume and frequency of antibiotic usage. Factors such as high disease prevalence, lack of government oversight, poor zoning regulations, and insufficient hygiene measures in animal husbandry further exacerbate this issue [10, 13]. Some of the most commonly used antibiotics in agriculture and livestock farming (including cattle production) include oxytetracycline, streptomycin, penicillin, oxalinic acid, gentamycin, tetracycline, ceftiofur, enrofloxacin, tulathromycin, florfenicol, phenicol, lacosamide, tilmacoxib, pleuromutilin, polypeptide, macrolide, carbadox, streptogramin, and bambermycin, among others [14]. Additionally, integrated agriculture–aquaculture systems, where aquaculture is supported by waste from both livestock and humans, pose a heightened risk of exposing animals, humans, and the environment to antibiotic residues [15]. Many farmers rely on local medicine vendors for guidance on antibiotic administration, which may further contribute to inappropriate usage. Wealthier farmers, having greater financial means, often use multiple antibiotics, potentially increasing antimicrobial resistance risks [16–18].

The term "pharmaceuticals" has been used in recent years to refer to a wide range of substances that are not currently subject to environmental regulations and whose effects on the environment and public health are still unknown [19, 20]. A variety of pharmaceutically active molecules, including steroids, antidepressants, antibiotics, steroids, antipyretics, stimulants, cosmetics, antimicrobials, and disinfectants, are commonly used on a regular basis, which results in the release of drug residues into various aquatic ecosystems worldwide [21]. Antibiotics and pharmaceutical residues are increasingly acknowledged as new water pollutants. Pseudo-persistent characteristics and the risks these drug residues provide to aquatic environments when discharged in significant amounts are the subject of recent research [22]. Due to their widespread manufacturing and consumption worldwide, pharmaceutically active chemicals are regularly released into aquatic environment [23]. Antibiotics, which are classified as chemotherapeutic substances or drugs that inhibit bacterial growth and multiplication in the host, are the most widely utilized pharmaceutically active substances [24]. Additionally, antibiotics are significantly increasing both human and animal life expectancy. It has been stated that people and animals utilize more than 250 antibiotic-active compounds worldwide, with China being the world's largest producer and consumer of antibiotics [25]. Global antibiotic usage is between 100 and 200 thousand metric tons/year, according to Ding et al. [26] and Akhter et al. [27]. According to Pan and Chu [28], antibiotics are used like honey for the health of both humans and animals. According to Qiao et al. [29], antibiotics have been used for prophylaxis, meta-prophylaxis, disease therapy, and growth promotion in healthy animals. The effects of antibiotics discharged into the aquatic environment on the health of people and animals are seriously questioned [20].

A significant portion of antibiotic residues have also been found in the marine environment. The majority of which are discharged through riverine discharges, sewage treatment facilities (STPs), and industrial effluents. Antibiotic traces have been discovered in soil, wastewater, sewage sludge, and animal feces. According to Huang et al. [30] and Sanusi et al. [31], antibiotics are excreted in various forms and in significant amounts as unchanged metabolites from the urine and excreta (10–90%) of living organisms. For instance, in countries including China, Korea, Europe, Canada, the United States, and Japan, reports of the common antiepileptic drug carbamazepine (CAB) in ground water, drinking water, sewage effluents, and river water have been made [32]. These antibiotics can affect on-target species (plants, animals, or microorganisms) once they are released into the environment. They can also accumulate these residues, even at low concentrations, and have physiological effects on plant, animal, or microbial populations, communities, and ecosystems [33]. Drug residue contamination in aquatic environments has recently drawn a lot of interest due to its unknown or unanticipated detrimental effects on human and aquatic health [34].

11.2 Fundamentals of Electrochemical Sensing

Electrochemical biosensors are analytical tools utilized in a number of domains, such as industrial and medical diagnostics, environmental monitoring, and food safety. These sensors are primarily made up of four parts: (a) a recognition element that interacts with the target molecule selectively; (b) a transducer that transforms the signal produced when the target interacts with the recognition element into a measurable output; (c) an electrolyte, which is the medium in which the electrochemical reaction occurs; and (d) electrodes, which are separated into reference, counter, and working electrodes and can be made of various materials such as carbon, gold, or other conductive materials. There are currently several varieties of electrochemical biosensors, such as impedimetric, potentiometric, and amperometric biosensors. Electrochemical signals produce a measurable change in impedance (electrochemical impedance spectroscopy, EIS), a measurable current (amperometry), a measurable potential or charge accumulation (potentiometry), or a measurably altered conductive property of a medium (conductometry) between electrodes [35, 36].

11.3 Types of Electrochemical Sensors

11.3.1 Amperometry Biosensors

In amperometry, an electroanalytical method, an indicator working electrode is subjected to a constant reducing or oxidizing potential, and the current generated is then measured. The magnitude of the observed current is typically determined by the concentration of the reduced or oxidized species. In the case of biosensors, the concentration of the analyte being studied is directly related to the reduction or oxidation of an electroactive molecule [37]. Amperometric biosensors, a type of electrochemical biosensors, enable the measurement of an analyte inside a sample matrix in this scenario by converting surface-level interactions caused by an electroactive species into a current signal [38]. Because of the intrinsic simplicity of the transducer, these biosensors can be employed with low-cost portable devices [39]. An amperometric transducer is used to study the charge transfer at phase interfaces, often between two electrodes that are separated by an electrolyte. These systems are called electrochemical cells, and they are used to describe the system in which the reaction takes place. One of the half-cells in the electrochemical compartment is closely monitored in order to examine the charge transfer occurring at the other half-cell, also known as the working electrode. Numerous amperometric methods, including square wave voltammetry (SWV), differential pulse voltammetry (DPV), and cyclic voltammetry (CV), can be used in electrochemical biosensors; the latter two are used in most commercial products (e.g., glucose tests) [38].

11.3.2 Potentiometric Biosensors

Potentiometry is an electrochemical method that measures the potential difference between an indicator electrode and a reference electrode inside an electrolyte to provide information regarding bioreceptor–analyte interactions, much to amperometry [40]. This approach has several benefits over amperometry, including lower power consumption and portability and compact size. Variations in analyte concentration are associated with possible changes in potentiometric biosensors at zero current. Concentration determines the electromotive force (EMF), commonly known as a cell's voltage and a measurement of the energy provided by a power source per unit charge. This relationship can be related using the Nernst equation (Eq. (11.1)) [41].

$$E_{\text{cell}} = E^{0}{}_{\text{cell}} - RT\,zF\ln Q \tag{11.1}$$

Here, E_{cell} stands for the cell's standard electrode potential, expressed in volts, R for the universal gas constant, T for temperature in Kelvin, z for ion charge (moles of electrons), F for Faraday's constant, and Q for reaction quotient.

11.3.3 Coulometric Biosensors

Coulometric analysis uses an applied current or a voltage to totally change an analyte from one oxidation state to another at the working electrode. Through direct or indirect measurement of the total current conveyed, this method determines the number of electrons moved across the electrochemical cell. Consequently, the concentration of the analyte can be ascertained by applying Faraday's law, which is based on the number of electrons passed (Eq. (11.2)) [42].

$$Q = nFNa \tag{11.2}$$

where F is Faraday's constant, Na is the analyte's moles, and n is the analyte's electron count per mole. This equation holds true when 100% of the applied current is used to change the analyte's oxidation state since the analyte's conversion rate when passing through the electrodes is nearly 100%, making minor current losses negligible [43]. One advantage of coulometric analysis over other techniques is that it is an absolute method that does not require calibration curves or chemical standards to measure the analyte [43].

11.3.4 Impedimetric Biosensors

After the target–recognition site interaction, an analytical technique known as electrochemical impedance spectroscopy (EIS) analyzes the impedance or resistance of a system in Ohms [43]. EIS has several advantages over other electrochemical sensing techniques since it can use small signal analysis to examine sinusoidal signal relaxations over a wide frequency range, from approximately 0.1 Hz to 1 MHz [44]. EIS can be used to analyze charge, mass, and diffusion transfer processes. Furthermore, EIS can examine underlying material properties or current operations that impact the conductance, resistance, or capacitance of a cell. Impedance and resistance are analyzed differently because impedance does not obey Ohm's law in a cell that receives a direct current (DC), whereas resistance does. This is due to the fact that while impedance only resists AC flow, resistance resists both DC and AC flow [45]. To calculate complex impedance, one can vary the excitation frequency of the applied potential over a range of frequencies. This is the sum of the system's real impedance, or resistance, which measures the resistance to a flowing current, and imaginary impedance, or opposition to a change in current [46].

11.4 Nanomaterials for Electrochemical Sensing of Antibiotics

Nanomaterials are regarded as the most appealing choice for the design and construction of "functional" transducers for electrochemical biosensing systems because of their distinct physical and chemical characteristics. These systems adaptability in terms of post-modification or functionalization has significantly enhanced analytical performance for antibiotic detection. Given these developments, the following section discusses the different nanomaterials used for the electrochemical detection of antibiotics.

11.4.1 Carbon-Based Electrodes for Antibiotics Detection

Carbon is abundant in the universe. Electrodes are made of carbon materials as graphene, Glassy carbon electrode (GCE), carbon nanotubes (CNTs), SPCE, boron-doped diamond (BDD), and others. Carbon-based electrodes are advantageous for electrochemical study in the following areas: adsorption, electron transfer kinetics, stability and reproducibility, low background current, and wide potential range limits [47]. First, a wide potential range limit and low background current can be observed thanks to its redox reactions on the carbon surface. It suggests that people's perceptions of carbon are diverse. Second are stability and productivity. The production of carbon materials is an example of producibility. Applying it to different analytical results, electrode areas, background currents, and electron transfer rates in the interim demonstrates stability. Third, the type of carbon substance has a significant impact on the electron transmission kinetics in carbon-based materials. It has been shown that carbon-based surfaces may efficiently adsorb both organic and inorganic substances. An earlier article on the practical application of carbon base electrodes was published by Dhanjai et al. Carbon-base electrodes are widely used in point-of-care, clinical, and biological diagnostics, as their study shows. In support of this assertion, they reviewed small biomolecules such as metabolites [48, 49], amino acids [50, 51], lipids [52], vitamins [53, 54], and big molecules such as proteins [55], enzymes, and nucleic acids [56]. Different pulse voltammetry (DPV), cyclic voltammetry (CV), and square wave voltammetry (SWV) were among the electrochemical techniques used. A low limit detection value between nM (nano mole) and μM (micro mole) is shown in the result [57, 58].

11.5 Carbon Nanotubes

One of the most effective nanostructured carbonaceous materials for biosensing applications, carbon nanotubes have garnered a lot of attention since their discovery. In addition to offering scaffolding for the effective immobilization of biomolecules, carbon nanotubes have a special set of physical, chemical, and electrical characteristics that make them ideal for identifying metabolites, analytes, and disease biomarkers [59]. Furthermore, by enhancing their functional and physical characteristics (such as solubility), the availability of several functionalization procedures allowed them to reach their full potential. Functionalized carbon nanotubes are therefore now required for the easy creation of a variety of nanodevices [60]. The nanostructuring of electrodes using carbon nanotubes (CNTs) or CNT-composites for electrochemical biosensors for antibiotic detection has been the subject of several investigations. With a lower limit of detection and a broader detection range for routinely used antibiotics, these nanomaterials produced a notable improvement in electrochemical performance [61]. It was suggested that levofloxacin be found in human urine samples using vertically aligned CNT-based electrochemical sensors [62]. Regarding levofloxacin in urine, the sensor displayed a detection limit of 75.2 nM over a detection range of 1–10 μM. This indicates that MWCNTs oriented vertically ought to have outstanding electrocatalytic activity when it comes to levofloxacin electrooxidation.

11.6 Graphene-Based Nanomaterials

The 2D carbon nanomaterial graphene has garnered a lot of interest from the sensor and electronics industries. Individual single-walled carbon nanotubes are said to have 60 times the conductivity, stability, sensitivity, and surface area of a single graphene nanosheet [63]. Because of its high specific surface area and efficient electron transmission, graphene is a desirable option for antimicrobial electrochemical sensing [64]. Tetracycline detection was investigated using screen-printed carbon electrodes enhanced with graphene due to its exceptional electrocatalytic activity [65]. The sensor showed promise for highly accurate tetracycline detection at 0.08 μM levels in milk and honey samples. Recently, 3-D reduced graphene oxide architectures were developed using a template-free and environmentally friendly method [66]. In comparison to the current metal/polymer based electrochemical sensors previously described for CAP detection, it was shown that the very porous nature of 3-DRGO/GCE imparted a stronger electrochemical response (LOD of 0.15 μM) [67]. Additionally, a three-dimensional nanostructured composite of graphene, carbon nitride, and Au was created to increase electrical conductivity and active sites in order to determine ciprofloxacin and chloramphenicol simultaneously [63]. With a Dl of 27 nM, these authors reported an excellent electrochemical response to CAP when ciprofloxacin was present. With respectable recoveries ranging from 95.0 to 101.2%, the sensor was also shown to be promising for the quantitative measurement of chloramphenicol in milk samples.

11.7 Magnetic Nanomaterials

The intrinsic characteristics of magnetic nanoparticles (MNs), including their vast surface area, physico-chemical properties, controllable size, and low production cost, have led to their use in a variety of biosensing applications [68]. Consequently, MNs with suitable physico-chemistry and customized surface characteristics have been actively incorporated into the transducer as immobilization supports for biorecognition components (such as aptamers), Molecular imprinted polymers (MIPs), peptides, and antibodies) to enable quick and precise detection of target analyte molecules. Additionally, because MNs are easily chemically modified, new multifunctional nanocomposites for a new generation of electrochemical sensors have been developed [69]. Novel electroanalytical techniques have been studied for the detection of antibiotics in a variety of matrices (e.g., human fluids, environmental, food, and beverage samples) due to the enormous potential of MNs. In order to create electrochemical immunosensors for a variety of antibiotics, antibodies were first coupled with MNs (as a recognition element). In one such study [70], a label-free electrochemical immunosensor for kanamycin detection was built using nanocomposites of magnetic nanoparticles with silver metal nanoparticles and thionine functionalized graphene sheets (TH-GS/GA/Ag@Fe_3O_4NPs). A modest DL (0.03 nM over a linear range of 0.1–33 nM) toward kanamycin was found by thorough SWV investigations. These scientists asserted that the high electrical conductivity of Ag @Fe_3O_4 nanocomposites and the good electron transfer ability of graphene nanosheets worked in concert to provide the exceptional electroanalytical performance. Excellent sensitivity, high stability, and good selectivity (such as for glucose, vitamin C, gentamicin, and neomycin) were demonstrated by the sensor, guaranteeing its use for kanamycin detection in food samples generated from animals. Similarly, to create a low-cost label-free electrochemical immunosensor for TET, carboxyl Fe_3O_4 nanoparticles were added to a gold electrode (using chitosan as a linker) [71].

11.8 Conclusions

Going forward, the incorporation of electrochemical-based devices into the sensing domain represents a noteworthy technical breakthrough, propelled by their adaptability and ease of use. Known for their many uses, these gadgets are now legitimate substitutes in fields including medical diagnosis and environmental monitoring. By offering precise and dependable measures that are essential for making well-informed decisions and maintaining public safety, they address some of the primary problems with the industry's gold-standard technology. Actually, the primary benefit of electrochemical sensors is their rapid and effective delivery of exact data. For example, these instruments may identify contaminants at extremely low levels in environmental monitoring, allowing for prompt actions to stop ecological harm. Electrochemical sensors are employed in medical diagnostics not just to identify antibiotics but also to track blood sugar levels, identify illness biomarkers, and even evaluate the effectiveness of therapies in real time. Because it enables better condition management and lessens the need for frequent hospital visits, this feature is especially advantageous for patients who need continuous monitoring. The sensitivity, cost, and usefulness of various electrochemical biosensor types should all be taken into account when comparing them. Amperometric sensors, for instance, are perfect for real-time monitoring because of their great sensitivity and quick response times. However, because they require specific electrodes, their production might be more costly. Compared to one another, potentiometric sensors may have lower sensitivity but are typically more affordable. Although they offer a cost-sensitivity balance, conductometric sensors—which monitor variations in electrical conductivity—may need more involved calibration techniques. Adoption and marketing of electrochemically based gadgets are not without difficulties, despite their many advantages. Even though these sensors are typically less expensive than many conventional gold-standard technologies, some consumers may find the initial outlay to be unaffordable. Additionally, continued developments in sensor technology should further increase accessibility and reduce prices. It is anticipated that advancements in materials science, such as the use of more reasonably priced and environmentally friendly materials for sensor construction, would further reduce the cost of these devices. Better manufacturing techniques can also increase the scalability of sensor production, which will result in more affordable and widely available products. The expensive price may prevent these sensors from being widely used, especially in environments with limited resources where financial limitations are a major issue. Thus, although electrochemical sensors'affordability is a benefit, more work is required to lower costs and improve the dependability and accessibility of these technologies for a wider range of users.

References

1. Sharma A, Istamboulie G, Hayat A, Catanante G, Bhand S, Marty JL (2017) Disposable and portable aptamer functionalized impedimetric sensor for detection of kanamycin residue in milk sample. Sensors Actuators B Chem 245:507–515. https://doi.org/10.1016/j.snb.2017.02.002
2. Ben Y, Caixia F, Min H, Liu L, Wong MH, Zheng C (2019) Human health risk assessment of antibiotic resistance associated with antibiotic residues in the environment: a review. Environ Res 169:483–493. https://doi.org/10.1016/j.envres.2018.11.040
3. You Y, Silbergeld EK (2014) Learning from agriculture: understanding low-dose antimicrobials as drivers of resistome expansion. Front Microbiol 5:284. https://doi.org/10.3389/fmicb.2014.00284
4. Hong P-Y, Al-Jassim N, Ansari MI, Mackie RI (2013) Environmental and public health implications of water reuse: antibiotics, antibiotic resistant bacteria, and antibiotic resistance genes. Antibiotics 2(3):367–399. https://doi.org/10.3390/antibiotics2030367
5. Hao H, Cheng G, Iqbal Z, Ai X, Hafiz H, Huang L, Dai M, Wang Y, Liu Z, Yuan Z (2014) Benefits and risks of antimicrobial use in food-producing animals. Front Microbiol 5:288. https://doi.org/10.3389/fmicb.2014.00288/full
6. Menkem ZE, Ngangom BL, Tamunjoh SSA, Boyom FF (2019) Antibiotic residues in food animals: public health concern. Acta Ecol Sin 39(5):411–415. https://doi.org/10.1016/j.chnaes.2018.10.004
7. Boeckel V, Thomas P, Brower C, Gilbert M, Grenfell BT, Levin SA, Robinson TP, Teillant A, Laxminarayan R (2015) Global trends in antimicrobial use in food animals. Proc Natl Acad Sci 112(18):5649–5654. https://doi.org/10.1073/pnas.1503141112
8. Bester LA, Essack SY (2012) Observational study of the prevalence and antibiotic resistance of Campylobacter spp. from different poultry production systems in KwaZulu-Natal, South Africa. J Food Prot 75(1):154–159. https://doi.org/10.4315/0362-028x_jfp-11-237
9. Maron DF, Smith TJS, Nachman KE (2013) Restrictions on antimicrobial use in food animal production: an international regulatory and economic survey. Glob Health 9:1–11. https://doi.org/10.1186/1744-8603-9-48
10. Islam KBMS, Shiraj-Um-Mahmuda S, Hazzaz-Bin-Kabir M (2016) Antibiotic usage patterns in selected broiler farms of Bangladesh and their public health implications. J Public Health Dev Countries 2(3):276–284
11. Kaplan J, Bowman M, Cordova C, Kar A (2014) Pharming chickens: it's time for the US poultry industry to demonstrate antibiotic stewardship. Natural Resources Defence Council, New York
12. Bashahun D, Odoch T (2015) Assessment of antibiotic usage in intensive poultry farms in Wakiso District, Uganda. Article 247. http://www.lrrd.org/lrrd27/12/bash27247.html
13. Mann A, Kiran Nehra JS, Rana, Dahiya T (2021) Antibiotic resistance in agriculture: perspectives on upcoming strategies to overcome upsurge in resistance. Curr Res Microb Sci 2:100030. https://doi.org/10.1016/j.crmicr.2021.100030

14. Kim DP, Saegerman C, Douny C, Dinh TV, Xuan BH, Binh Dang V, Hong NP, Scippo M-L (2013) First survey on the use of antibiotics in pig and poultry production in the red River Delta region of Vietnam. Food Public Health 3(5):247–256. https://doi.org/10.5923/j.fph.20130305.03
15. Braykov NP, Eisenberg JNS, Grossman M, Zhang L, Vasco K, Cevallos W, Muñoz D et al (2016) Antibiotic resistance in animal and environmental samples associated with small-scale poultry farming in northwestern Ecuador. Msphere 1(1):10–1128. https://doi.org/10.1128/msphere.00021-15
16. Sekyere JO (2014) Antibiotic types and handling practices in disease management among pig farms in Ashanti region, Ghana. J Vet Med 2014:531952. https://doi.org/10.1155/2014/531952
17. Kim J, Kang C-I, Joo E-J, Ha YE, Cho SY, Gwak G-Y, Chung DR, Peck KR, Song J-H (2014) Risk factor of community-onset spontaneous bacterial peritonitis caused by fluoroquinolone-resistant Escherichia coli in patients with cirrhosis. Liver Int 34(5):695–699. https://doi.org/10.1111/liv.12374
18. Adebowale OO, Adeyemo OK, Awoyomi O, Dada R, Adebowale O (2016) Antibiotic use and practices in commercial poultry laying hens in Ogun state Nigeria. Rev Elev Med Vet Pays Trop 69(1):41–45. https://doi.org/10.19182/remvt.31170
19. Bao Q, Wang Y, Tang S, Ye F, Zhiyang Y, Ye Q, Wang W (2022) Science of the total environment uptake and accumulation of erythromycin in leafy vegetable sand induced phytotoxicity and dietary risks. Sci Total Environ 830:154785. https://doi.org/10.1016/j.scitotenv.2022.154785
20. Gworek B, Kijeńska M, Wrzosek J, Graniewska M (2021) Pharmaceuticals in the soil and plant environment: a review. Water Air Soil. https://doi.org/10.1007/s11270-020-04954-8
21. Singh V, Suthar S (2021) Occurrence, seasonal variations, and ecological risk of pharmaceuticals and personal care products in river Ganges at two holy cities of India. Chemosphere 268:129331
22. Hossain A, Nakamichi S, Tani K (2018) Occurrence and ecological risk of pharmaceuticals in river surface water of Bangladesh. Environ Res 165(January):258–266. https://doi.org/10.1016/j.envres.2018.04.030
23. Biswas P, Vellanki BP (2021) Occurrence of emerging contaminants in highly anthropogenically influenced river Yamuna in India. Sci Total Environ 782:146741. https://doi.org/10.1016/j.scitotenv.2021.146741
24. Keerthanan S, Jayasinghe C, Biswas JK, Vithanage M (2021) Pharmaceutical and personal care products (PPCPs) in the environment: plant uptake, translocation, bioaccumulation, and human health risks. Crit Rev Environ Sci Technol 51(12):1221–1258. https://doi.org/10.1080/10643389.2020.1753634
25. Gu D, Feng Q, Guo C, Hou S, Lv J, Zhang Y, Yuan S, Zhao X (2019) Occurrence and risk assessment of antibiotics in manure, soil, wastewater, groundwater from livestock and poultry farms in Xuzhou, China. Bull Environ Contam Toxicol 103(4):590–596. https://doi.org/10.1007/s00128-019-02692-0
26. Ding H, Yixiao W, Zhang W, Zhong J, Lou Q, Yang P, Fang Y (2017) Occurrence, distribution, and risk assessment of antibiotics in the surface water of Poyang Lake, the largest freshwater lake in China. Chemosphere 184:137–147. https://doi.org/10.1016/j.chemosphere.2017.05.148
27. Akhter S, Bhat MA, Hashem A, Abd-Allah EF, Ahmed S, Siddiqi WA, Kulsoom I, Nisa FU (2023) Profiling of antibiotic residues in soil and vegetables irrigated using pharmaceutical-contaminated water in the Delhi stretch of the Yamuna River, India. Water. https://doi.org/10.3390/w15234197
28. Pan M, Chu LM (2017) Fate of antibiotics in soil and their uptake by edible crops. Sci Total Environ 599–600:500–512. https://doi.org/10.1016/j.scitotenv.2017.04.214
29. Zhou X, Wang J, Chao L, Liao Q, Gudda FO, Ling W (2020) Antibiotics in animal manure and manure-based fertilizers: occurrence and ecological risk assessment. Chemosphere 255:127006. https://doi.org/10.1016/j.chemosphere.2020.127006

30. Huang W, Qiu Q, Chen M, Shi J, Huang X, Kong Q, Long D, Chen Z, Yan S (2019) Determination of 18 antibiotics in urine using LC-QqQ-MS/MS. J Chromatogr B Analyt Technol Biomed Life Sci 1105:176–183. https://doi.org/10.1016/j.jchromb.2018.12.019
31. Sanusi IO, Olutona GO, Wawata IG, Onohuean H (2023) Occurrence, environmental impact and fate of pharmaceuticals in groundwater and surface water: a critical review. Environ Sci Pollut Res 30(39):90595–90614. https://doi.org/10.1007/s11356-023-28802-4
32. Shao Y, Wang Y, Yuan Y, Xie Y (2021) Science of the total environment a systematic review on antibiotics misuse in livestock and aquaculture and regulation implications in China. Sci Total Environ 798(18):149205. https://doi.org/10.1016/j.scitotenv.2021.149205
33. Tran NH, Hoang L, Nghiem LD, Nguyen NMH, Ngo HH, Guo W, Trinh QT, Mai NH, Chen H, Nguyen DD (2019) Occurrence and risk assessment of multiple classes of antibiotics in urban canals and lakes in Hanoi, Vietnam. Sci Total Environ 692:157–174
34. Zafar R, Bashir S, Nabi D, Arshad M (2021) Occurrence and quantification of prevalent antibiotics in wastewater samples from Rawalpindi and Islamabad, Pakistan. Sci Total Environ 764:142596. https://doi.org/10.1016/j.scitotenv.2020.142596
35. Grieshaber D, MacKenzie R, Vörös J, Reimhult E (2008) Electrochemical biosensors-sensor principles and architectures. Sensors 8(3):1400–1458. https://doi.org/10.3390/s80314000
36. Lopes LC, Santos A, Bueno PR (2022) An outlook on electrochemical approaches for molecular diagnostics assays and discussions on the limitations of miniaturized technologies for point-of-care devices. Sens Actuators Rep 4:100087. https://doi.org/10.1016/j.snr.2022.100087
37. Shoyiga HO, Fayemi OE (2025) Conductive polymers and derivatives as recognition element for electrochemical sensing of food and drug additives: a brief perspective. Heliyon 11(2):e41575J
38. Hammond JL, Formisano N, Estrela P, Carrara S, Tkac J (2016) Electrochemical biosensors and nanobiosensors. Essays Biochem 60(1):69–80. https://doi.org/10.1042/EBC20150008
39. Pedersen T, Fojan P, Pedersen AKN, Magnusson NE, Gurevich L (2023) Amperometric biosensor for quantitative measurement using Sandwich immunoassays. Biosensors 13(5):519. https://doi.org/10.3390/bios13050519
40. Walker NL, Roshkolaeva AB, Chapoval AI, Dick JE (2021) Recent advances in potentiometric biosensing. Curr Opin Electrochem 28:100735. https://doi.org/10.1016/j.coelec.2021.100735
41. Magar HS, Hassan RYA, Mulchandani A (2021) Electrochemical impedance spectroscopy (EIS): principles, construction, and biosensing applications. Sensors 21(19):6578. https://doi.org/10.3390/s21196578
42. Ziyatdinova G, Budnikov H (2021) Analytical capabilities of coulometric sensor systems in the antioxidants analysis. Chem 9(5):91. https://doi.org/10.3390/chemosensors9050091
43. Razem M, Ding Y, Morozova K, Mazzetto F, Scampicchio M (2022) Analysis of phenolic compounds in food by coulometric array detector: a review. Sensors 22(19):7498. https://doi.org/10.3390/s22197498
44. Lazanas A, Prodromidis MI (2023) Electrochemical impedance spectroscopy–a tutorial. ACS Measure Sci 3(3):162–193. https://doi.org/10.1021/acsmeasuresciau.2c00070
45. Macdonald JR (1987) Impedance spectroscopy and its use in analyzing the steady-state AC response of solid and liquid electrolytes. J Electroanal Chem Interfacial Electrochem 223(1-2):25–50. https://doi.org/10.1016/0022-0728(87)85249-X
46. Caldara M, Lowdon JW, van Wissen G, Ferrari AG-M, Crapnell RD, Cleij TJ, Diliën H, Banks CE, Eersels K, van Grinsven B (2023) Dipstick sensor based on molecularly imprinted polymer-coated screen-printed electrodes for the single-shot detection of glucose in urine samples—from fundamental study toward point-of-care application. Adv Mater Interf 10(18):2300182. https://doi.org/10.1002/admi.202300182
47. Uslu B, Ozkan SA (2007) Electroanalytical application of carbon based electrodes to the pharmaceuticals. Anal Lett 40(5):817–853. https://doi.org/10.1080/00032710701242121
48. Thirumalraj B, Palanisamy S, Chen S-M, Yang C-Y, Periakaruppan P, Lou B-S (2015) Direct electrochemistry of glucose oxidase and sensing of glucose at a glassy carbon electrode modi-

fied with a reduced graphene oxide/fullerene-C60 composite. RSC Adv 5(95):77651–77657. https://doi.org/10.1039/C5RA12018G
49. Li Z, Sheng L, Meng A, Xie C, Zhao K (2016) A glassy carbon electrode modified with a composite consisting of reduced graphene oxide, zinc oxide and silver nanoparticles in a chitosan matrix for studying the direct electron transfer of glucose oxidase and for enzymatic sensing of glucose. Microchim Acta 183:1625–1632. https://doi.org/10.1007/s00604-016-1791-x
50. Wang Y, Wang W, Li G, Liu Q, Wei T, Li B, Jiang C, Sun Y (2016) Electrochemical detection of L-cysteine using a glassy carbon electrode modified with a two-dimensional composite prepared from platinum and Fe_3O_4 nanoparticles on reduced graphene oxide. Microchim Acta 183:3221–3228. https://doi.org/10.1007/s00604-016-1974-5
51. Martín A, Batalla P, Hernández-Ferrer J, Martínez MT, Escarpa A (2015) Graphene oxide nanoribbon-based sensors for the simultaneous bio-electrochemical enantiomeric resolution and analysis of amino acid biomarkers. Biosens Bioelectron 68:163–167. https://doi.org/10.1016/j.bios.2014.12.030
52. Mondal K, Ali MA, Singh C, Sumana G, Malhotra BD, Sharma A (2017) Highly sensitive porous carbon and metal/carbon conducting nanofiber based enzymatic biosensors for triglyceride detection. Sensors Actuators B Chem 246:202–214. https://doi.org/10.1016/j.snb.2017.02.050
53. Sýs M, Švecová B, Švancara I, Metelka R (2017) Determination of vitamin E in margarines and edible oils using square wave anodic stripping voltammetry with a glassy carbon paste electrode. Food Chem 229:621–627. https://doi.org/10.1016/j.foodchem.2017.02.068
54. Hameed S, Munawar A, Khan WS, Mujahid A, Ihsan A, Rehman A, Ahmed I, Bajwa SZ (2017) Assessing manganese nanostructures based carbon nanotubes composite for the highly sensitive determination of vitamin C in pharmaceutical formulation. Biosens Bioelectron 89:822–828. https://doi.org/10.1016/j.bios.2016.10.005
55. Ruiyi L, Jiajia W, Ling L, Zaijun L (2016) Ultrasensitive direct detection of dsDNA using a glassy carbon electrode modified with thionin-functionalized multiple graphene aerogel and gold nanostars. Microchim Acta 183:1641–1649. https://doi.org/10.1007/s00604-016-1793-8
56. Miodek A, Mejri N, Gomgnimbou M, Sola C, Korri-Youssoufi H (2015) E-DNA sensor of mycobacterium tuberculosis based on electrochemical assembly of nanomaterials (MWCNTs/PPy/PAMAM). Anal Chem 87(18):9257–9264. https://doi.org/10.1021/acs.analchem.5b01761
57. Shamsipur M, Farz Serafín L, Agüí VL, Yáñez-Sedeño PINGARRÓN, Pingarrón JM (2014) Determination of prolactin hormone in serum and urine using an electrochemical immunosensor based on poly (pyrrolepropionic acid)/carbon nanotubes hybrid modified electrodes. Sens Actuators B Chem 195:494–499. https://doi.org/10.1016/j.snb.2014.01.055
58. Sinha A, Xianbo L, Lingxia W, Tan D, Li Y, Chen J, Jain R (2018) Voltammetric sensing of biomolecules at carbon based electrode interfaces: a review. TrAC Trends Anal Chem 98:174–189. https://doi.org/10.1016/j.trac.2017.11.010
59. Thiele-Bruhn S (2003) Erratum: Thiele-Bruhn (2003): pharmaceutical antibiotic compounds in soils: a review; issue 166 (2), 145-167. J Plant Nutr Soil Sci 166(4):546–546. https://doi.org/10.1002/jpln.200390066
60. Sun Y-P, Kefu F, Lin Y, Huang W (2002) Functionalized carbon nanotubes: properties and applications. Acc Chem Res 35(12):1096–1104. https://doi.org/10.1021/ar010160v
61. Gayen P, Chaplin BP (2016) Selective electrochemical detection of ciprofloxacin with a porous nafion/multiwalled carbon nanotube composite film electrode. ACS Appl Mater Interfaces 8(3):1615–1626. https://doi.org/10.1021/acsami.5b07337
62. Moraes FC, Silva TA, Cesarino I, Lanza MRV, Machado SAS (2013) Antibiotic detection in urine using electrochemical sensors based on vertically aligned carbon nanotubes. Electroanalysis 25(9):2092–2099. https://doi.org/10.1002/elan.201300261
63. Kuila T, Bose S, Khanra P, Mishra AK, Kim NH, Lee JH (2011) Recent advances in graphene-based biosensors. Biosens Bioelectron 26(12):4637–4648. https://doi.org/10.1016/j.bios.2011.05.039

64. Yuan Y, Zhang F, Wang H, Gao L, Wang Z (2018) A sensor based on au nanoparticles/carbon nitride/graphene composites for the detection of chloramphenicol and ciprofloxacin. ECS J Solid-State Sci Technol 7(12):M201–M208. https://doi.org/10.1149/2.0111812jss
65. Filik H, Avan AA, Aydar S, Ozyurt D, Demirata B (2016) Determination of tetracycline on the surface of a high-performance graphene modified screen-printed carbon electrode in milk and honey samples. Curr Nanosci 12(4):527–533
66. Zhang X, Zhang Y-C, Zhang J-W (2016) A highly selective electrochemical sensor for chloramphenicol based on three-dimensional reduced graphene oxide architectures. Talanta 161:567–573. https://doi.org/10.1016/j.talanta.2016.09.013
67. Karthik R, Govindasamy M, Chen S-M, Mani V, Lou B-S, Devasenathipathy R, Hou Y-S, Elangovan A (2016) Green synthesized gold nanoparticles decorated graphene oxide for sensitive determination of chloramphenicol in milk, powdered milk, honey and eye drops. J Colloid Interface Sci 475:46–56. https://doi.org/10.1016/j.jcis.2016.04.044
68. Akbarzadeh A, Samiei M, Davaran S (2012) Magnetic nanoparticles: preparation, physical properties, and applications in biomedicine. Nanoscale Res Lett 7:1–13. https://doi.org/10.1186/1556-276X-7-144
69. Rocha-Santos TAP (2014) Sensors and biosensors based on magnetic nanoparticles. TrAC Trends Anal Chem 62:28–36. https://doi.org/10.1016/j.trac.2014.06.016
70. Yu S, Wei Q, Bin D, Dan W, Li H, Yan L, Ma H, Zhang Y (2013) Label-free immunosensor for the detection of kanamycin using ag@ Fe3O4 nanoparticles and thionine mixed graphene sheet. Biosens Bioelectron 48:224–229. https://doi.org/10.1016/j.bios.2013.04.025
71. Liu X, Zheng S, Yuxin H, Li Z, Luo F, He Z (2016) Electrochemical immunosensor based on the chitosan-magnetic nanoparticles for detection of tetracycline. Food Anal Methods 9:2972–2978. https://doi.org/10.1007/s12161-016-0480-z

Chapter 12
Biocompatible Nanomaterials Based Sensors for Environmental and Biomedical Applications

Manjunath Megalamani, Jyothi Abbar, B. M. Santhosh, K. N. Shravana Kumara, N. Hareesha, D. A. Ayush Gowda, and Yuvarajgouda Patil

Abstract The biocompatible nanomaterials with smart sensor technology marks a transformative era in both environmental and biomedical monitoring. These biocompatible nanomaterials offer extraordinary and unique properties like high surface area, high aspect ratios, tunable electronic and optical properties, and minimal toxicity, making them ideal candidates for interfacing with biological systems and detecting trace levels of environmental contaminants. This chapter explores various classes of biocompatible nanomaterials-including carbon-based structures, metallic and metal-oxide nanoparticles and their integration into electrochemical sensing platforms. Applications environmental samples and biomedical samples are critically reviewed. The discussion extends to the physicochemical parameters and crystalline structures of nanomaterials, which critically influence their behavior and functional performance, synthesis strategies, biofunctionalization approaches, performance metrics, and real-world deployment challenges. Finally, the current challenges and future prospects of biocompatible nanomaterials for fostering their practical utilization in smart sensors are also addressed.

M. Megalamani
Department of Chemistry, Atria Institute of Technology, Bengaluru, India

Visvesvaraya Technological University, Belagavi, Karnataka, India

J. Abbar
BMS Institute of Technology and Management (Affiliated to Visvesvaraya Technological University), Bengaluru, Karnataka, India

B. M. Santhosh · K. N. Shravana Kumara · N. Hareesha · Y. Patil (✉)
Department of Chemistry, GM University, Davangere, Karnataka, India

D. A. Ayush Gowda
Graduate School of Biomedical Science and Engineering, Hanyang University, Seoul, South Korea

R. Naik et al. (eds.), *Biocompatible Nanomaterials for Smart Sensors*, Nanostructure Science and Technology,
https://doi.org/10.1007/978-3-032-01745-1_12

Keywords Biocompatible nanomaterials · Smart sensors · Electrochemical sensors · Synthesis · Environmental monitoring · Biomedical applications

12.1 Introduction

Significant environmental damage has resulted from the world's rapid industrialization and urbanization, calling for creative methods of management and monitoring. For example, the widespread use of agrochemicals, such as fertilizers and pesticides, during the Green Revolution in India has contaminated soil and water, compromising human health and agricultural output [1]. The persistence of these chemicals in the environment has also led to bioaccumulation and biomagnification in food chains, further exacerbating ecological and health issues [2]. While industrial growth remains a major driver of development, the Earth is concurrently grappling with numerous environmental crises, including habitat loss, climate change, and ecosystem degradation on a global scale [3]. Urban sprawl, deforestation, and increasing greenhouse gas emissions have intensified the pressure on natural ecosystems [4]. Developing powerful environmental monitoring systems to identify, examine, and remove dangerous contaminants is vital to restoring ecological balance and mitigating associated effects [5]. Advanced monitoring technologies, combined with machine learning and big data analytics, are becoming integral to early detection and prevention strategies [6]. In this context, sensor technologies have significantly enhanced everyday life, with applications spanning healthcare, pharmacy, environmental monitoring, and industrial processes [7]. Sensors, which detect changes in their environment or source, capture signals, and trigger appropriate responses, have evolved rapidly to address growing demands for sensitivity, reliability, and cost-effectiveness [8].

The development of innovative sensor materials, such as conductive polymers and hybrid organic-inorganic compounds, has further enhanced sensor performance [9]. Biosensors, in particular, have emerged as critical tools for early disease diagnosis, rapid detection of diverse drugs and chemicals, and long-term physiological monitoring [10]. Their biocompatibility ensures safe interaction with biological systems—an essential feature for clinical and wearable applications. Advanced materials like hydrogels and self-healing polymers have improved sensor lifespan and user comfort, extending their utility in healthcare [11]. In the healthcare sector, these devices enable real-time monitoring, accurate data collection, and personalized medical interventions. This is especially crucial given the increasing prevalence of chronic illnesses, infectious disease outbreaks, and lifestyle-related health issues [12]. Wearable technologies, integrated with smart sensors, are transforming traditional healthcare into proactive, preventive models. These sensors can track vital physiological parameters such as heart rate, blood glucose, and body temperature, as well as external factors like ambient humidity and physical activity [13]. Continuous data capture and analysis facilitate early disease detection, improving the accuracy and scope of health monitoring and enabling timely interventions [14].

Electrochemical sensors have proven particularly effective in quantifying samples of different origins. They allow rapid analysis with minimal or no sample treatment, offering low detection limits and operational simplicity [15]. Compared to optical methods, electrochemical techniques require less expensive instrumentation, making them accessible for various applications [16]. Recent advances in nanostructured electrodes, such as carbon nanotubes and graphene, have further enhanced their sensitivity and reliability [17].

High-performance smart sensors require materials that are not only biocompatible but also stable and adaptable to diverse conditions. Nanoparticles and nanocomposites have emerged as ideal candidates due to their tuneable physicochemical properties and biodegradability [18]. These nanomaterials provide robust biological support in sensor systems while ensuring consistent performance across varying physiological and environmental conditions. Additionally, their integration into sensor technologies has revolutionized wearable health monitoring and environmental sensing [19]. Nanomaterial-based sensors, with high surface-to-volume ratios and excellent electrical conductivity, can detect environmental parameters like air and water quality as well as physiological signals such as biomarkers [20]. Two-dimensional materials like MoS_2 and black phosphorus are advancing the fabrication of ultra-sensitive wearable sensors [21]. Features such as stretchability, flexibility, and breathability make these devices ideal for continuous, noninvasive health monitoring [22]. Scalable production techniques like inkjet printing are further driving the adoption of wearable sensor technologies [23].

The environmental and health impacts of nanomaterials remain areas of active investigation. To ensure the sustainable and safe development of these advanced materials, it is critical to evaluate factors influencing their toxicity and biocompatibility comprehensively [24]. Nanotoxicology studies focusing on particle size, shape, and surface chemistry are vital for risk assessment and safe design [25]. This chapter explores the biocompatibility, synthesis strategies, biofunctionalization techniques, and performance metrics of nanomaterials used in sensors. Additionally, it addresses limitations and outlines future directions for integrating biocompatible nanomaterials into smart sensor systems [26].

12.2 Overview of Nanomaterials

Nanomaterials differ significantly from their bulk counterparts, primarily due to their nanoscale dimensions (1–100 nm), which endow them with enhanced mechanical strength, increased surface reactivity, and novel optical and electrical behaviors. They are categorized into various forms including 0D nanoparticles, 1D nanowires, 2D sheets like graphene, and 3D nanostructures [27]. Compositionally, they span a broad range from carbon-based and metal-based nanomaterials to metal oxides, organic, and composite materials. Figure 12.1 [28] showcases the different forms of nanoparticles such as polymeric nanoparticles (PNPs), magnetic nanoparticles

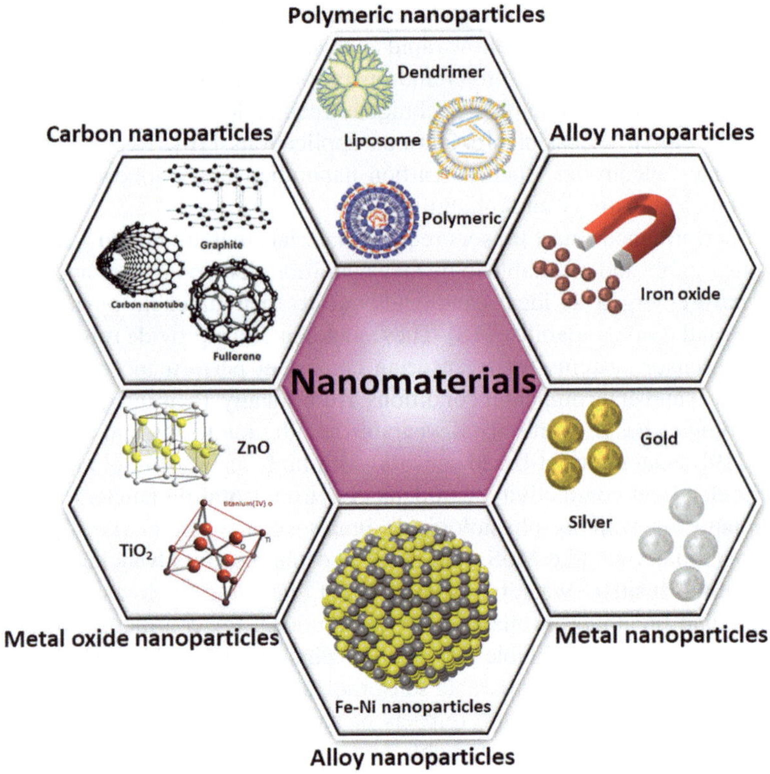

Fig. 12.1 Different forms of nanoparticles

(MNPs), metal and metal oxide nanoparticles, alloy nanoparticles, and carbon nanomaterials.

These nanomaterials are being widely explored across diverse industries such as healthcare [29], electronics, energy [30], and environmental remediation. In wearable biosensing technologies, their high surface-to-volume ratio, electrical conductivity, flexibility, and biocompatibility contribute to the development of multifunctional, high-performance sensors capable of continuous, real-time health monitoring. As research advances, the intersection of nanotechnology and biocompatibility will continue to drive innovation in both medical and environmental applications [31], ensuring safety, sustainability, and improved quality of life.

12.3 Biocompatible Nanomaterials

Biocompatibility is a fundamental property that governs the safe and effective application of nanomaterials in biomedical and environmental fields. As the integration of nanotechnology into healthcare continues to expand, ensuring the compatibility of nanomaterials with biological systems has become critically important. This involves an understanding of how nanomaterials interact with cellular components such as proteins, lipids, and carbohydrates, which in turn influences cellular functions and systemic biological responses [32].

To ensure biocompatibility, various strategies and technologies have been developed. These include surface modification to enhance stability and reduce toxicity, biomimetic designs that mimic natural biological structures, and rigorous in vitro and in vivo testing. In addition, the field of nanotoxicology and evolving regulatory frameworks play vital roles in establishing safety guidelines for clinical and environmental use [33].

12.3.1 Synthesis of Biocompatible Nanoparticles (BNMs)

Biocompatible nanomaterials (BNMs) can be synthesized using various physical, chemical, and biological techniques (Fig. 12.2), each offering unique advantages in controlling nanoparticle structure, morphology, and functionality. Among the commonly employed methods, coprecipitation is widely used due to its simplicity, where nanoparticle formation occurs through simultaneous nucleation and growth; however, it often faces challenges related to product purity, reproducibility, and limited applicability to uncharged species [34]. Hydrothermal synthesis, which involves high-temperature and high-pressure reactions in sealed autoclaves, allows precise control over crystal growth and morphology but is hindered by high operational costs and limited reaction monitoring [35]. Alternatively, microwave-assisted synthesis utilizes rapid and selective heating, resulting in higher yields, improved purity, and scalability compared to conventional heating methods [36]. Another approach, microemulsion, relies on reactant exchange within nanodroplets, enabling spontaneous nanoparticle formation with high stability; however, the process is sensitive to pH and temperature and requires substantial surfactant use [37]. Ultrasonic (Sono - chemical) synthesis leverages ultrasonic cavitation to generate extreme reaction conditions, facilitating rapid nanoparticle formation with controlled morphologies under ambient conditions, though its scalability remains a challenge due to energy intensity [38]. Meanwhile, template-based synthesis involves confining nanoparticles within porous host structures, ensuring shape and size control while imparting multifunctionality to the resulting materials [39]. In alignment with green chemistry principles, biological (green) synthesis employs plant extracts, microbes, or other biomolecules as reducing and stabilizing agents, offering eco-friendly and sustainable nanoparticle production, albeit with limitations such as slower reaction

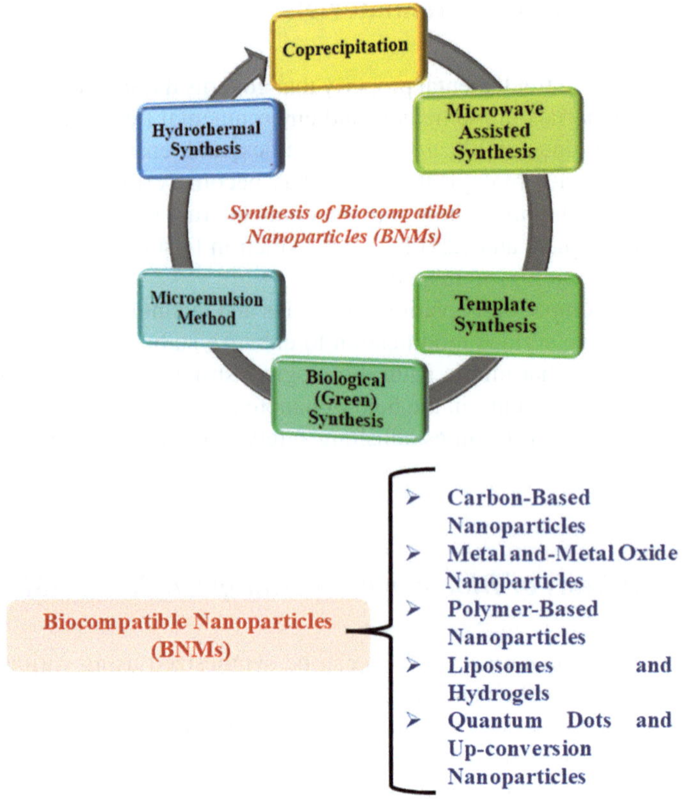

Fig. 12.2 Various techniques for preparation of biocompatible nanoparticles

rates and polydispersity in particle sizes [40]. Collectively, these synthesis methods provide diverse pathways for producing BNMs tailored for environmental and biomedical applications while addressing the need for biocompatibility and functional versatility.

12.3.2 Types of Biocompatible Nanoparticles

Biologically relevant nanomaterials (BNMs) are engineered for safe interaction with biological and environmental systems and have been widely explored for sensing and remediation of environmental and biomedical samples. Their morphology critically influences their properties, with structures such as nanowires and nanotubes promoting electron delocalization, while shapes like nanotriangles and nanotubes enhance charge accumulation at sharp edges. Star-shaped and tetrapod nanoparticles further exhibit superior catalytic activity [41]. These features make

BNMs ideal for designing highly sensitive and selective electrochemical sensors [42]. To enhance performance, conventional electrodes are often modified with advanced nanomaterials like graphene, graphene quantum dots, carbon nanotubes, conductive polymers, and metal or metal oxide nanoparticles (Table 12.1).

Carbon-Based BNMs are among the most prominent, including carbon nanodots, graphene oxide (GO), and carbon nanotubes (CNTs). These materials possess excellent physicochemical properties, such as high surface area, superior electrical conductivity, and remarkable stability, making them highly effective for electrochemical and optical sensing applications. GO, particularly when functionalized with metal oxides like FeOOH or Mn_3O_4, exhibits enhanced electrocatalytic activity and adsorption capacity, making it highly suitable for water purification systems [43]. CNTs, both single-walled and multi-walled, are also widely used for their strong adsorption potential and conductivity, although their recovery post-treatment and potential environmental persistence remain challenging [44].

Metal and metal oxide nanoparticles, including those of iron, gold, silver, and manganese, are also heavily employed in both sensing and remediation. These nanoparticles serve as active agents in redox reactions and exhibit unique optical properties that are exploited in colorimetric and fluorescence-based sensors—particularly for detecting fluoride ions via mechanisms such as fluorescence resonance energy transfer (FRET) [45].

Polymer-based BNMs, such as chitosan and its derivatives, have been extensively studied for their excellent biocompatibility and high affinity for anionic pollutants. Chitosan, when functionalized with transition metals like zirconium and lanthanum, significantly enhances the adsorption capacity for fluoride and nitrate removal [46, 47]. Synthetic polymers like polydimethylsiloxane and polypyrrole are also utilized as matrices in biosensor designs for enzyme immobilization, improving both sensitivity and durability of detection platforms [48].

Table 12.1 Different types of biocompatible nanoparticles and there uses

Type	Examples	Main use	Key features
Carbon-based	CNTs, GO, C-dots	Sensing and adsorption	High conductivity, large surface area
Metal/metal oxide	Fe_3O_4, AuNPs, Mn_3O_4	Remediation sensing sensors	Catalytic, magnetic, optical properties
Polymer-based	Chitosan, PEDOT	Adsorption enzyme support	Biocompatible, functionalizable
Quantum dots	CdTe, CdSe/ZnS	Fluorescent sensing	High sensitivity, ratiometric detection
UCNPs	$NaYF_4$:Yb,Er	Fluorescent sensing	NIR excitation, low background
MOFs	La-MOFs	Adsorption and sensing	Porous, tunable chemistry
Natural nanomaterials	Tea dregs, avocado seed chars	Adsorption	Low-cost, green synthesis

Quantum dots (QDs)—such as CdTe and CdSe/ZnS—offer superior fluorescence characteristics and are frequently used in turn-on/turn-off or ratiometric fluorescent sensors for ultra-sensitive detection of fluoride ions [49]. Upconversion nanoparticles (UCNPs), which convert near-infrared (NIR) light to visible emission, provide deeper penetration and reduced background interference, enhancing sensor accuracy [50].

Metal-organic frameworks (MOFs), characterized by their high porosity and tunable surface chemistry, have demonstrated remarkable potential in selective adsorption and sensing of fluoride and nitrate ions. However, limitations such as pH sensitivity and component leaching still restrict their practical application [50]. Lastly, natural or bio-derived nanoparticles, synthesized from agricultural and food waste such as tea dregs, coffee grounds, and avocado seeds, offer a cost-effective and eco-friendly alternative. These materials are particularly promising for use in low-resource settings, although their adsorption efficiency may not always match that of engineered nanomaterials [51, 52]. Despite their diversity and proven lab-scale effectiveness, scaling up the application of BNMs for real-time and industrial wastewater treatment remains an ongoing research focus.

12.4 Environmental Applications

Rapid industrialization and technological progress have significantly impacted the environment, leading to increased pollution of air, water, and soil. Contaminants such as heavy metals, excess nutrients, microorganisms, and organic pollutants pose serious threats to public health and ecosystems, contributing to declining air and water quality, soil degradation, and biodiversity loss. To effectively address these challenges, advanced tools are needed to accurately detect and quantify these harmful substances. Electrochemical sensors have emerged as a promising solution for environmental monitoring due to their low cost, simplicity, and portability.

12.4.1 Heavy Metal Detection

Metal sensing has become a crucial area of research because of the detrimental effects of heavy metal ions on living organisms and ecosystems. Manami et al. [53] developed a ZnO/reduced graphene oxide (ZnO/RGO) nanocomposite-based electrochemical sensor for the sensitive and selective detection of Pb (II) ions. The sensor exhibited excellent performance under optimized conditions, achieving a detection limit of 7 ppb and a quantification limit of 21 ppb. The developed sensor demonstrated high reproducibility, stability, and effective regeneration, along with outstanding selectivity for Pb(II) over other metal ions due to engineered recognition sites within the composite. Practical applicability was validated through the analysis of environmental water samples, yielding quantitative recoveries ranging from 98.8% to 102.1%. Similarly, Chen et al. [54] developed a 3D melamine-doped reduced graphene oxide/MXene composite aerogel (3D-MGMA) integrated with a screen-printed carbon electrode (SPCE) for the simultaneous and sensitive detection of Cd^{2+}, Pb^{2+}, and Zn^{2+} ions in environmental samples. The aerogel was synthesized via a self-assembly process involving L-cysteine, graphene oxide (GO), and melamine/MXene, achieving uniform dispersion of melamine/MXene within the 3D network. The 3D porous rGO structure enhanced electrical conductivity, while melamine-doped MXene introduced amino functionalities that promoted efficient metal ion complexation. Stability was confirmed over ten consecutive tests, with low relative standard deviations (RSDs) of 0.82% for Cd^{2+}, 0.63% for Zn^{2+}, and 0.66% for Pb^{2+}. Reproducibility and repeatability were validated using five independent sensors across 100–900 μg/L, showing consistent peak currents and linear differential pulse anodic stripping voltammetry (DPASV) signals with RSDs below 5%, confirming excellent sensor performance.

12.4.2 *Pesticides and Herbicides Sensing*

The use of pesticides and herbicides plays a vital role in modern agriculture; however, the persistence of their residues poses serious health risks. To address this, various sensor technologies have been developed for the effective detection of pesticide residues in real-world samples. Sivaji et al. [55] reported the development of a highly sensitive and rapid electrochemical sensor for detecting carbendazim (CBZ), utilizing a glassy carbon electrode modified with an indium oxide–carbon nanofiber composite (In_2O_3@CNF/GCE) (Fig. 12.3). The In_2O_3@CNF composite was extensively characterized to confirm its morphology, elemental composition, and crystallinity. Electrochemical impedance spectroscopy (EIS) revealed a low charge transfer resistance (Rct) of 103 Ω, indicating enhanced electron transport across the electrode interface. Both differential pulse voltammetry (DPV) and cyclic voltammetry (CV) demonstrated excellent electrocatalytic activity toward CBZ. The sensor displayed a broad linear detection range from 1.0×10^{-8} to 2.3×10^{-4} M, an impressive detection limit of 4.2 nM, and a high sensitivity of 13.6 $\mu A\ \mu M^{-1}\ cm^{-2}$. It also exhibited excellent reproducibility, repeatability, cyclic stability, and long-term operational stability. The sensor's practical applicability was validated through successful CBZ recovery (98.43–99.8%) from real fruit samples such as apples and tangerines, highlighting its potential for reliable CBZ monitoring in agricultural products. In another study, electrochemical detection of atrazine (ATZ) in tap and wastewater samples was achieved using NiO nanoparticles decorated MnO_2 nanotubes synthesized via a hydrothermal method [56]. The sensor demonstrated a detection limit of 400 nM and a linear detection range from 0.8 μM to 120 μM.

Additionally, Traiwatcharanon et al. [57] recently developed a screen-printed electrode modified with copper oxide nanoparticles for the detection of a combined herbicide formulation of butachlor and propanil (commercially known as Daramox).

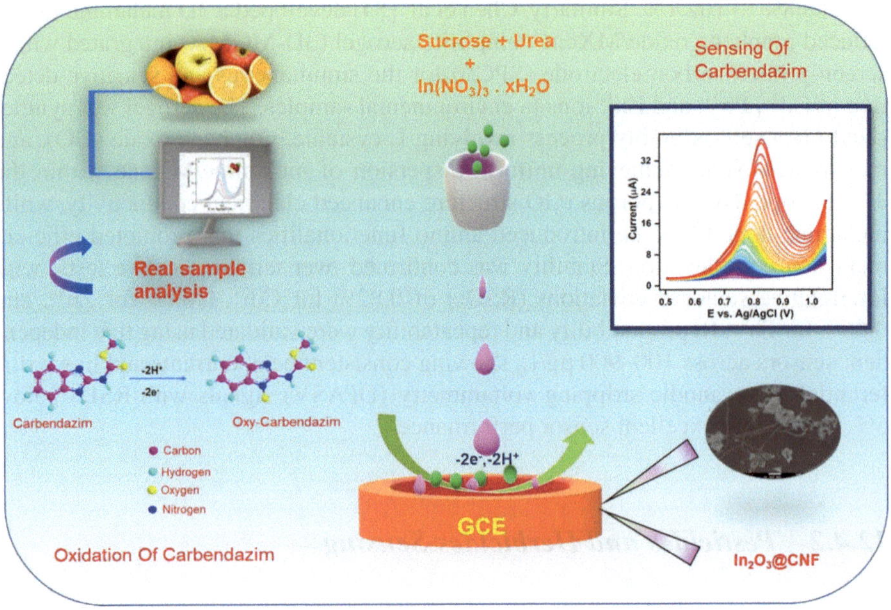

Fig. 12.3 Electrochemical sensor for detecting carbendazim (CBZ), utilizing a glassy carbon electrode modified with an indium oxide–carbon nanofiber composite (In_2O_3@CNF/GCE). Adapted with permission from [55], 2025 Elsevier Publisher

The sensor achieved a sensitivity of 5.32 μA mM^{-1} cm^{-2}, a linear range of 1–5 mM, and a detection limit of 0.05 mM under ambient conditions.

12.4.3 Hydrogen Peroxide (H_2O_2) Detection

Hydrogen peroxide (H_2O_2) is a widely utilized green oxidizing agent across various industries, including textiles, paper manufacturing, environmental protection, and healthcare. It is also commonly employed as a bleaching and sterilizing agent in the food industry. However, due to its strong oxidative properties, excessive exposure or residual levels of H_2O_2 can lead to oxidative stress, potentially damaging cellular components such as proteins, nucleic acids, and membranes. This can increase the risk of chronic diseases and other health complications [58].

To address the need for sensitive detection of H_2O_2, Yao et al. [59] developed a novel porous palladium (Pd) nanostructure using a soft micellar templating method involving TMB (Tetramethylbenzidine) and the surfactant CTAB (Cetyltrimethylammonium Bromide). The resulting Pd nanoparticles, averaging 2 nm in size, were uniformly distributed within the template. The constructed electrochemical sensor demonstrated high sensitivity for H_2O_2 detection, with a detection limit of 0.25 μmol/L and a broad linear range of 0.25–900 μmol/L. It was successfully applied to detect H_2O_2 released by MDA-MB-435 cells at an operating potential of −0.34 V.

Further innovations in H_2O_2 sensor development include the work of Ding et al. [60] who synthesized MnO nanoparticles supported on carbon nanofibers (CNFs), enhancing the availability of active sites and boosting electrocatalytic performance. Similarly, Balasubramanian reported a sensor based on Mn_2CuO_4 (MCO) microspheres fabricated via a solvothermal approach, capable of detecting cellular H_2O_2 in the presence of CHAPS. This sensor was applied to monitor H_2O_2 secretion from A375 cells, highlighting its potential for biomedical applications. These advanced H_2O_2 sensors exhibit outstanding performance due to the synergistic interactions among their composite materials.

12.4.4 Other Toxic Compounds

Other toxic compounds, such as CO, formaldehyde, aflatoxin B1, or those containing phenolic groups, are introduced into the environment through various channels including agriculture, industrial activities, sewage drainage, and pharmaceutical drugs [61]. Nevertheless, these compounds are toxic because of their mutagenic, carcinogenic, and/or teratogenic characteristics, thereby presenting significant health concerns.

Researcher Huang et al. [62] developed a highly sensitive semiconductor gas sensor for formaldehyde (HCHO) detection using bimetallic Ag nanoparticle (Ag NP) and Au nanocage (Au NC) co-modified In_2O_3 (Ag_6Au_1/In_2O_3) synthesized via a one-step microwave-assisted solvothermal method (Fig. 12.4). The sensor exhibited a remarkable response of 277 toward 5 ppm HCHO at a low operating temperature of 170 °C—approximately 7, 38, and 102 times higher than 6 wt.% Ag/In_2O_3, 1 wt.% Au/In_2O_3, and pristine In_2O_3, respectively. With a detection limit as low as 26 ppb and enhanced selectivity, the Ag_6Au_1/In_2O_3-based sensor demonstrated exceptional performance for trace-level HCHO monitoring in indoor air and human breath. The synergistic catalytic effect of Ag and Au played a key role in improving sensitivity and efficiency. First-principles density functional theory (DFT) calculations were employed to elucidate the underlying sensing mechanism, highlighting the potential of bimetallic modification for next-generation gas sensing technologies.

Panda et al. [63] developed a reduced graphene oxide (rGO)-based gas sensor for selective carbon monoxide (CO) detection. Synthesized via a modified Hummers' method, the rGO offered high conductivity and surface area. The sensor showed ~71% sensitivity to 30 ppm CO at room temperature, with fast response/recovery (<30 s for 10 ppm) and excellent selectivity over other reducing gases. A biosynthesized gold nanoparticle-modified glassy carbon electrode (AuNPs/GCE) was developed by Shi et al. [64] for the sensitive detection of nitrite (NO_2^-). The Au nanoparticles were synthesized using *Lycoris radiata* pollen extract, chosen for its

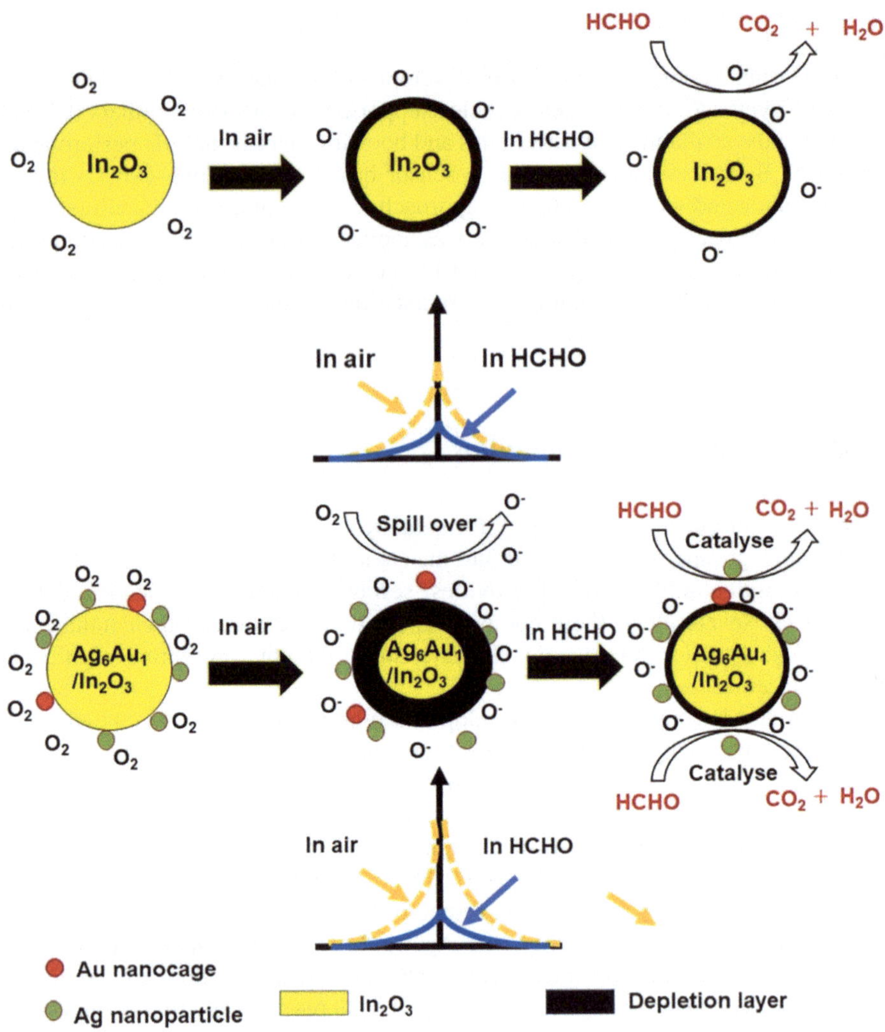

Fig. 12.4 The sensing mechanism diagram of formaldehyde on the surface of In_2O_3 sensor and Ag_6Au_1/In_2O_3 sensor. Adapted with permission from [62], 2022 Elsevier Publisher

high flavone content, which facilitated uniform nanoparticle formation and prevented agglomeration, yielding a conductive and stable electrode interface. The AuNPs/GCE sensor exhibited a linear response for NO_2^- in the range of 0.01–3.8 mM, achieving a detection limit of 2.4 μM. The sensor demonstrated excellent anti-interference performance and was successfully applied for NO_2^- determination in drinking water, confirming its practical applicability for environmental monitoring.

12.5 Biomedical Application

The pharmaceutical and biomedical sectors are currently facing significant analytical challenges, necessitating the development of innovative and reliable detection methods. In this context, the creation of advanced sensor technologies has become increasingly crucial. Among the diverse analytical targets, pharmaceutical formulations and biomedical matrices present some of the most complex systems due to their intricate compositions. As a result, designing analytical systems that are both highly sensitive and selective remains a demanding task. Affinity-based sensors, in particular, offer promising solutions by streamlining analytical procedures and minimizing overall costs.

12.5.1 Glucose Sensor

In the human body, glucose maintenance is very important. The improper absorption of glucose results in diabetes, which may lead to renal and neural problems. Hence, recognition of glucose is medically significant for the detection of diabetes in the human body. For the analysis of glucose, Balasubramanian, et al. [65] developed a non-enzymatic glucose sensor based on ilmenite-type $NiMnO_3$ (NMO) nanostructures synthesized via a simple hydrothermal method. Surface morphology analysis revealed that the NMO formed interconnected ultrathin nanosheets (~15 nm), which enhanced ion diffusion and improved electrocatalytic activity. The fabricated sensor exhibited excellent electrocatalytic performance for glucose detection, with a broad dynamic range (0.00005–1.01 mM and 1.01–13.2 mM), a low detection limit of 14 nM, and a high sensitivity of 7560.8 $\mu A \cdot mM^{-1} \cdot cm^{-2}$. Practical applicability was demonstrated by accurate glucose detection in human serum, saliva, and sweat, showing reliable recoveries compared to a commercial glucose sensor. Zalke et al. [66] developed a cost-effective, eco-friendly screen-printed chemiresistive sensor on glossy photo paper for enzymatic glucose detection. Graphene-based ink formed low-resistance interdigitated electrodes, functionalized with polyaniline, glucose oxidase, and green-synthesized silver nanoparticles. The sensor showed a linear response from 198 nM to 30 mM glucose, with a detection limit of 198 nM and high sensitivity (291.19 $\mu A\ mM^{-1}\ cm^{-2}$), making it suitable for point-of-care diagnostics. Recently, Dai et al. [67] developed a non-enzymatic electrochemical glucose sensor using a ZnO-modified nitrogen-doped porous carbon (ZnO/N-PC-10) composite. Synthesized via freeze-drying, carbonization, and hydrothermal methods, the sensor showed excellent glucose oxidation at 0.60 V with a wide linear range (2–3280 μM), high sensitivity (255.99 $\mu A\ mM^{-1}\ cm^{-2}$), and a low detection limit (0.39 μM). Its enhanced performance was due to the synergistic effect of N-doped carbon and nano-ZnO, and it demonstrated strong stability, selectivity, and clinical applicability.

12.5.2 *Pharmaceutical Drug Analysis*

The rapid expansion of the pharmaceutical industry, driven by global population growth, has brought new biomedical and environmental challenges. Beyond ensuring quality in drug production and delivery, there is a growing demand for accurate, sensitive, portable, and affordable technologies to detect patient overdosing and monitor pharmaceutical contaminants in water sources. In this context, nanomaterial-based electrochemical sensors and biosensors have gained significant attention for their high sensitivity, selectivity, real-time detection capabilities, and user-friendly operation [68].

Patil et al. [69] developed a Ce-doped ZnO/reduced graphene oxide (Ce-ZnO/r-GO) nanocomposite-modified glassy carbon electrode (GCE) for the sensitive and selective detection of vonoprazan (VPN) (Fig. 12.5), a potassium-competitive acid blocker used in peptic ulcer treatment. The nanocomposite was synthesized via a conventional hydrothermal method and thoroughly characterized for its morphological, structural, and electrochemical properties. Under optimized conditions, the modified electrode demonstrated a distinct voltammetric response toward VPN oxidation, with a detection limit of 1.8×10^{-8} M and a linear range of 2.0×10^{-7} to 8.0×10^{-6} M. The charge transfer rate constant (ks) was calculated as 1.17×10^{-6} cm s^{-1}, and the electroactive surface area was estimated to be 0.081 cm^2. The sensor exhibited excellent selectivity, stability, and reproducibility, and was successfully applied to the analysis of VPN in human serum and urine samples, achieving favorable recovery rates. Megalamani et al. [70] developed a voltammetric sensor using a YSZ/MoS_2-modified carbon paste electrode for sensitive chlorzoxazone (CXZ) detection. Electrochemical studies using EIS, CV, and SWV revealed a clear oxidation peak at pH 7.0, with SWV enabling a detection limit of 1.8×10^{-9} M. The sensor showed high sensitivity and was effectively applied to

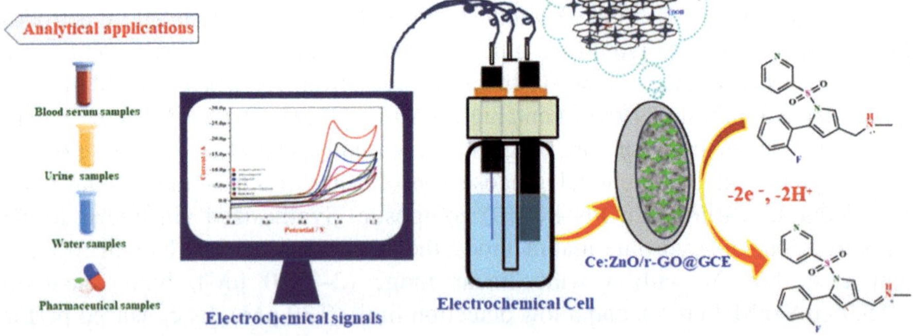

Fig. 12.5 Ce-doped ZnO/reduced graphene oxide (Ce-ZnO/r-GO) nanocomposite-modified glassy carbon electrode (GCE) for the sensitive detection of vonoprazan (VPN). Adapted with permission from [68], 2024 Springer Publisher

pharmaceutical, biological, and environmental samples. Additionally, El-Zahry et al. [71] developed a Mo@GO/CS-V_2O_5 nanorod-modified electrochemical sensor for simultaneous detection of diclofenac (DCF) and posaconazole (POS) in spiked biological samples. The bimetallic nanostructures with GO and carbon spheres enhanced charge transfer and sensitivity. The sensor showed excellent peak separation, high recovery, and low RSDs in urine and plasma, highlighting its suitability for pharmaceutical quality control. Recently, Yazdanian et al. [72] developed a Fe_3O_4-MXene nanocomposite sensor for sensitive electrochemical detection of idarubicin (IDA). The hydrothermally synthesized nanocomposite showed uniform Fe_3O_4 distribution (22–44 nm) and a crystallite size of 37 nm. It exhibited enhanced electrocatalytic activity with a detection range of 0.1–4 μM and a detection limit of 26.43 nM. The sensor demonstrated high selectivity, repeatability, and was effectively applied to plasma and urine analysis.

12.5.3 Detections of Pathogens

The rapid and accurate detection of pathogens like *Mycobacterium tuberculosis*, *Staphylococcus aureus*, and *Salmonella typhimurium* remains a significant challenge in clinical microbiology. Traditional methods, including culture-based techniques, nucleic acid detection, and immunological assays, have been widely employed. However, despite advancements, these approaches often remain time-consuming and labor-intensive. Consequently, there is a critical need for the development of innovative analytical sensors that offer rapid, sensitive, and specific pathogen identification. Ismardi et al. [73] successfully developed a prototype sensor for the detection of *E. coli* by modifying a screen-printed carbon electrode (SPCE) with a ZnO/PVA nanocomposite. The ZnO/PVA nanocomposite material was characterized using FESEM (Field Emission Scanning Electron Microscope) and TEM (Transmission Electron Microscope), revealing average ZnO particle sizes of 38 nm and 59 nm, respectively, with the difference attributed to the arrangement of particles during FESEM analysis. XRD analysis confirmed the presence of both ZnO and PVA, with an average ZnO crystal size of 34 nm. The sensor exhibited high sensitivity, as demonstrated by the I-V test, and stable electrochemical performance over 30 cycles in cyclic voltammetry (CV) testing. The sensor's electrochemical response was found to vary with increasing bacterial concentration, and the limit of detection (LOD) was determined to be 4.41×10^7 CFU, with a rapid detection time of approximately 3–5 min, significantly faster than conventional methods. The promising results and potential for further integration of biosensing components suggest that the ZnO/PVA nanocomposite-based electrochemical sensor could be a leading option for the rapid detection of *E. coli*.

Recently, Ismardi et al. [74] developed a high-performance ratiometric electrochemical aptasensor based on graphene quantum dots and copper-based metal-organic frameworks (GQDs/Cu-MOF) for the sensitive detection of *Staphylococcus aureus* (*S. aureus*) (Fig. 12.6). The sensor exhibited outstanding sensitivity

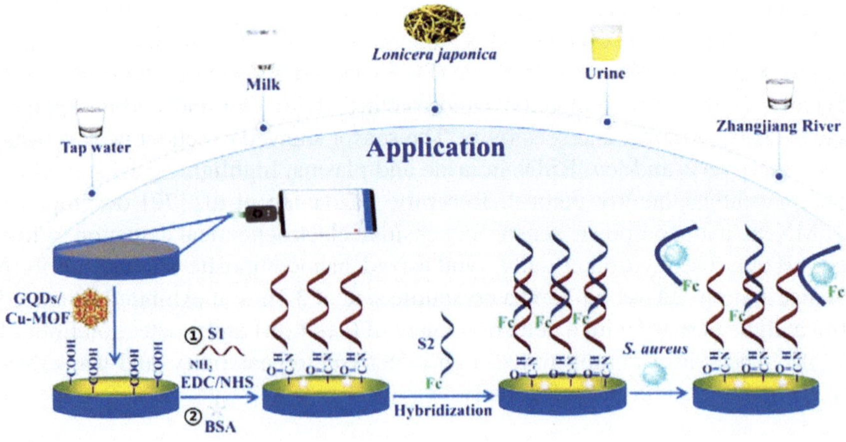

Fig. 12.6 Electrochemical aptasensor based on GQDs/Cu-MOF nanocomposite for the on-site determination of *Staphylococcus aureus* in tap water, milk, *Lonicera japonica* urine, and Zhangjiang River. Adapted with permission from [74], 2025 Elsevier Publisher

(0.97 CFU·mL^{-1}), a broad linear detection range (5.0 × 10 to 5.0 × 10^{8} CFU·mL^{-1}), and excellent specificity, stability, and performance in real sample analysis. Leveraging the programmable dynamic range and structural versatility of aptamer design, the aptasensor enabled rapid threshold-level detection. This universal sensing strategy also holds potential for detecting other foodborne pathogens such as *E. coli* O157:H7, *B. cereus*, *Y. enterocolitica*, and *L. monocytogenes*, offering a promising platform for food safety monitoring and early disease warning systems.

12.6 Current Challenges

Despite several research and advantages, electrochemical sensors utilizing biocompatible nanomaterials still face certain significant challenges as discussed below.

12.6.1 Biocompatibility and Long-Term Stability

Electrochemical sensors often operate in dynamic biological or environmental matrices, where long-term stability is a key concern. Despite biocompatible nanomaterials such as graphene oxide (GO) or gold nanoparticles (AuNPs) show very good initial performance, their interaction with biological fluids (e.g., blood, sweat, or saliva) can lead to surface modifications, degradation, or leaching of metal ions. Over time, this can reduce sensor performance with respect to the sensitivity and accuracy. For biomedical sensing, the biocompatibility of nanomaterials must be validated for prolonged exposure, especially if it is an in vivo application.

12.6.2 Surface Fouling and Signal Drift

In both environmental and biomedical sample analysis, the electrochemical sensors are prone to the surface accumulating interfering substances such as proteins, ions, or particulate matter. This phenomenon can be called as surface fouling and it can alter the sensor's surface properties and lead to signal drift or diminished performance. Fouling is a problematic issue in electrochemical sensors if used in the human body or complex environmental systems, where contamination is obvious. Developing effective anti-fouling approaches or novel and improved surface modification strategies are very essential to prevent the sensor surface degradation.

12.6.3 Reproducibility and Batch-to-Batch Variability

The synthesis of nanomaterials is often a batch process, and inconsistencies in size, morphology, and functionalization can lead to significant variability in sensor performance. Achieving reproducibility in large-scale production still remains a challenge. Variations in the physical and chemical properties of nanomaterials may happen during their synthesis and it can lead to the fluctuations in the sensor response, and thereby it can complicate the development of futuristic sensors for commercial viability and in clinical applications.

12.7 Future Scope of Biocompatible Nanomaterials in Smart Electrochemical Sensors

The integration of biocompatible nanomaterials into electrochemical sensors is paving a way for next-generation sensing technologies which is promising, and several areas have shown a great potential. As we are mostly moving toward miniaturized devices, wearable, implantable and flexible sensors, biocompatible nanomaterials are promising materials due to their biocompatibility and flexibility. As the technology is driving toward artificial intelligence and machine learning they can be applied in the predictive detection and early detection of any abnormalities by analyzing real -time data from these sensors.

In future, electrochemical sensors can be converted into self-powered and portable devices for continuous monitoring of environmental sources and thereby eliminating the need for external power supplies. Sensors can be made with biocompatible inks incorporated within nanomaterials for environmental and biomedical diagnosis.

In addition, sensors —preferably electrochemical sensors —must be build with the capacity of multi-analyte detection, which can be applied for simultaneous detection of several biomarkers or environmental samples with high precision and selectivity. Especially multi-analyte detection can help in biomedical diagnosis where in the human sample may not be collected each time for each test. This will reduce the need for frequent pricking of blood from patients and help in the detection of several parameters from a single sample. Finally, the approaches wherein green synthesis and biodegradable materials are used will ensure environmental safety. Moreover, multiplexed detection and 3D-printed platforms will allow rapid, personalized solutions for disease diagnosis and environmental tracking. Together, these innovations will open a new era of intelligent, responsive, and sustainable sensing technologies with incorporation of nano and bioactive materials [75, 76].

12.8 Conclusions

The synergistic combination of nanotechnology and advanced sensor technologies provides a powerful framework for tackling urgent environmental and biomedical challenges. The examples discussed showcase the remarkable potential of biocompatible nanomaterials in crafting highly sensitive, selective, and efficient sensors for a wide spectrum of analytes. Offering significant advantages over conventional techniques, nanomaterial-based sensors enable superior detection of environmental pollutants (heavy metals, pesticides, toxic gases), biological markers (glucose, pharmaceutical drugs), and pathogens. Their distinctive physicochemical characteristics, coupled with innovative synthesis and functionalization, result in enhanced performance, portability, and cost-effectiveness. While research continues to evaluate and mitigate potential risks, the progress presented underscores the transformative impact of these materials on developing next-generation monitoring systems for improved health and sustainability. The ongoing convergence of materials science, nanotechnology, electrochemistry, and biology promises even more advanced and widely applicable smart sensor technologies. The incorporation of biocompatible nanomaterials into sensing platforms, particularly electrochemical sensors, represents a transformative strategy essential for addressing critical challenges in environmental monitoring and biomedical diagnostics. These advanced materials amplify sensor capabilities through their exceptional surface area, conductivity, and biological compatibility, enabling sensitive, selective, and real-time detection across a broad range of analytes.

References

1. Sharma M (2022) Environmental degradation and industrialization: a review. J Environ Stud 127:199–210
2. Aktar MW, Sengupta D, Chowdhury A (2009) Impact of pesticide use in agriculture: their benefits and hazards. Interdiscip Toxicol 2(1):1–12. https://doi.org/10.2478/v10102-009-0001-7
3. Intergovernmental Panel on Climate Change (IPCC) (2021) Climate change 2021: the physical science basis. Cambridge University Press, Cambridge. https://doi.org/10.1017/9781009157896
4. Grayson M (2019) Big data analytics and environmental monitoring: a review. Environ Monit Assess 191(9):556. https://doi.org/10.1007/s10661-019-7735-8
5. Javaid M, Haleem A (2019) Sensors for daily life: a comprehensive overview. Mater Today Proc 17:477–483. https://doi.org/10.1016/j.matpr.2019.06.477
6. Chen M, Ma Y, Li Y, Wu D, Zhang Y (2017) Smart clothing: connecting human with clouds and big data for sustainable health monitoring. Mob Netw Appl 21(5):825–845. https://doi.org/10.1007/s11036-016-0745-1
7. Torsi L, Farinola GM, Marinelli F, Tanese MC, Omar OH, Valli L, Ciccarella G (2008) Organic field-effect transistor sensors: a tutorial review. Chem Soc Rev 37(2):302–313. https://doi.org/10.1039/B703045C
8. Turner APF (2013) Biosensors: sense and sensitivity. Science 290(5495):1315–1317. https://doi.org/10.1126/science.290.5495.1315
9. Zang Y, Zhang F, Di C, Zhu D (2015) Advances of flexible pressure sensors toward artificial intelligence and health care applications. Mater Horizons 2(2):140–156. https://doi.org/10.1039/C4MH00147G
10. Justino CI, Rocha-Santos TA, Duarte AC (2010) Review of analytical figures of merit of sensors and biosensors in clinical applications. TrAC Trends Anal Chem 29(10):1172–1183. https://doi.org/10.1016/j.trac.2010.07.008
11. Yuk H, Lu B, Zhao X (2019) Hydrogel bioelectronics. Chem Soc Rev 48(6):1642–1667. https://doi.org/10.1039/C8CS00595H
12. Lee H, Song C, Hong YS, Kim MS, Cho HR, Kang T, Lee H (2017) Wearable/disposable sweat-based glucose monitoring device with multistage transdermal drug delivery module. Sci Adv 3(3):e1601314. https://doi.org/10.1126/sciadv.1601314
13. Heikenfeld J, Jajack A, Rogers J, Gutruf P, Tian L, Pan T, Kim J (2018) Wearable sensors: modalities, challenges, and prospects. Lab Chip 18(2):217–248. https://doi.org/10.1039/C7LC00914C
14. Bandodkar AJ, Wang J (2014) Non-invasive wearable electrochemical sensors: a review. TrAC Trends Anal Chem 76:173–185. https://doi.org/10.1016/j.trac.2015.11.016
15. Wang J (2006) Electrochemical biosensors: towards point-of-care cancer diagnostics. Biosens Bioelectron 21(10):1887–1892. https://doi.org/10.1016/j.bios.2005.10.027
16. Scida K, Stege PW, Haby G, Messina GA, García CD (2011) Recent applications of carbon-based nanomaterials in analytical chemistry: critical review. Anal Chim Acta 691(1–2):6–17. https://doi.org/10.1016/j.aca.2011.01.027
17. Pumera M (2010) Electrochemical properties of carbon nanotubes and graphene for nanomedicine. Sensors 10(5):4753–4770. https://doi.org/10.3390/s100504753
18. Wang X, Liu Z, Zhang T (2017) Flexible sensing electronics for wearable/attachable health monitoring. Small 13(25):1602790. https://doi.org/10.1002/smll.201602790
19. Xue G, Zhang W, Xu Z, Gao J (2021) Nanomaterials for wearable biosensors. Adv Funct Mater 31(6):2007683. https://doi.org/10.1002/adfm.202007683
20. Yao S, Zhu Y (2014) Wearable multifunctional sensors using printed stretchable conductors made of silver nanowires. Nanoscale 6(4):2345–2352. https://doi.org/10.1039/C3NR05496A
21. Akinwande D, Brennan CJ, Bunch JS, Egberts P, Felts JR, Gao H, Zhu Y (2017) A review on mechanics and mechanical properties of 2D materials—graphene and beyond. Extreme Mech Lett 13:42–77. https://doi.org/10.1016/j.eml.2017.01.008

22. Lipomi DJ, Bao Z (2011) Stretchable, elastic materials and devices for solar energy conversion. Energy Environ Sci 4(9):3314–3328. https://doi.org/10.1039/C1EE01730E
23. Derby B (2010) Inkjet printing of functional and structural materials: fluid property requirements, feature stability, and resolution. Annu Rev Mater Res 40:395–414. https://doi.org/10.1146/annurev-matsci-070909-104502
24. Fadeel B, Bussy C, Merino S, Vázquez E, Flahaut E, Mouchet F, Kostarelos K (2018) Safety assessment of graphene-based materials: focus on human health and the environment. ACS Nano 12(11):10582–10620. https://doi.org/10.1021/acsnano.8b04758
25. Nel A, Xia T, Mädler L, Li N (2006) Toxic potential of materials at the nanolevel. Science 311(5761):622–627. https://doi.org/10.1126/science.1114397
26. Bhattacharya K, Mukherjee SP, Gallud A, Burkert SC, Bistarelli S, Bellucci S, Fadeel B (2016) Biological interactions of carbon-based nanomaterials: from coronation to degradation. Nanomedicine 12(2):333–351. https://doi.org/10.1016/j.nano.2015.09.005
27. Byakodi M, Shrikrishna NS, Sharma R, Bhansali S, Mishra Y, Kaushik A, Gandhi S (2022) Emerging 0D, 1D, 2D, and 3D nanostructures for efficient point-of-care biosensing. Biosens Bioelectron X 12:100284
28. Syama S, Mohanan PV (2018) The promising biomedical applications of engineered nanomaterials. In: Handbook of nanomaterials for industrial applications. Elsevier, Amsterdam, pp 530–542. https://doi.org/10.1016/B978-0-12-813351-4.00030-4
29. Pushpanjali PA, Manjunatha JG, Hareesha N, D'Souza ES, Charithra MM, Prinith NS (2021) Voltammetric analysis of antihistamine drug cetirizine and paracetamol at poly (L-leucine) layered carbon nanotube paste electrode. Surf Interfaces 24:101154
30. Cheng TM, Hsieh CY, Cao YC, Kongvarhodom C, Saukani M, Yougbaré S, Lin LY (2025) Novel design of nickel and vanadium sulfide/oxide composites with tunable metal ratios as efficient energy storage materials. J Energy Storage 107:114971
31. Sabet M, Sabet S (2025) Biocompatible graphene polymer composites: new Frontiers in biomedical and environmental applications. Polym Plast Technol Mater 69:1–33
32. Jagtap P, Raut R, Gade A, Dave S (2025) Green synthesis of nanomaterials from bio-waste for efficient photocatalytic sensors. In: Bio-waste-derived carbon materials and their applications, especially as sensors. Elsevier, Amsterdam, pp 137–153
33. Aminnezhad S, Hama NH, Hasan AH, Bagheri F, Alavi M (2025) Applications of biocompatible polymeric nanomaterials in three-dimensional (3D) scaffolds: bacterial infections and diabetes. Int J Biol Macromol 301:140331
34. Bharath G, Rambabu K, Banat F, Hai A, Arangadi AFAF, Ponpandian N (2019) Enhanced electrochemical performances of peanut shell derived activated carbon and its Fe_3O_4 nanocomposites for capacitive deionization of Cr(VI) ions. Sci Total Environ 691:691–703
35. Chikan V, McLaurin EJ (2016) Rapid nanoparticle synthesis by magnetic and microwave heating. Nano 6(5):85
36. Ranjha MMAN, Shafique B, Rehman A, Mehmood A, Ali A, Zahra SM et al (2022) Biocompatible nanomaterials in food science, technology, and nutrient drug delivery: recent developments and applications. Front Nutr 8:1141
37. Salabat A, Mirhoseini F (2018) A novel and simple microemulsion method for synthesis of biocompatible functionalized gold nanoparticles. J Mol Liq 268:849–853
38. Samrot AV, Sahithya CS, Selvarani J, Purayil SK, Ponnaiah P (2021) A review on synthesis, characterization and potential biological applications of superparamagnetic iron oxide nanoparticles. Curr Res Green Sustain Chem 4:100042
39. Shen L, Qiao Y, Guo Y, Meng S, Yang G, Wu M (2014) Facile co-precipitation synthesis of shape-controlled magnetite nanoparticles. Ceram Int 40(1):1519–1524
40. Vallejo-Domínguez D, Rubio-Rosas E, Aguila-Almanza E, Hernández-Cocoletzi H, Ramos-Cassellis ME, Luna-Guevara ML, Rambabu K (2022) Ultrasound in the deproteinization process for chitin and chitosan production. Ultrason Sonochem 72:105417
41. Baig N, Sajid M, Saleh TA (2019) Recent trends in nanomaterial-modified electrodes for electroanalytical applications. TrAC Trends Anal Chem 111:47–61

42. Aarti DP, Basavaraj RB, Abbar JC, Reddy MM, Jisha PK (2025) High-performance green light-emitting CPE modified BaZrO3: Tb3+ nanopigments: a potential material for WLEDs and electrochemical sensing applications. J Ind Eng Chem 149:645–656
43. Liu Y, Wang J, Zheng H, Wang A, Zhang J (2015) Graphene oxide anchored with akaganéite for fluoride removal: effects of organic ligands and pH. J Environ Chem Eng 3(3):1628–1636
44. Zarei A, Pourzamani H, Hassani AH (2013) Removal of fluoride from drinking water using carbon nanotubes coated with nano-alumina. Desalin Water Treat 51(7–9):1707–1714
45. Wen D, Zhang K, Li Y, Han J (2021) Efficient fluoride removal using a novel nanomagnetite-graphite-lanthanum composite. Chemosphere 278:130484
46. Zhang Z, Shen W, Pan L (2013) Removal of fluoride from aqueous solution by chitosan-based adsorbents: a review. Chem Eng J 226:306–314
47. Prabhu SM, Rambabu K, Bharath G, Banat F (2021) Defluoridation of water using graphene oxide and nano-hydroxyapatite composites. Int J Environ Stud 78(5):734–747
48. Tyagi S, Rawtani D, Khatri N, Tharmavaram M (2018) Strategies for nitrate removal from aqueous environment using nanotechnology: a review. J Water Process Eng 21:84–95
49. Yang Y, Wu C, Li X, Zhang R, Zhu Y (2020) Fluorescent quantum dots for fluoride detection: a review of recent developments. J Photochem Photobiol B Biol 203:111745
50. Yin C, Huang Q, Zhu G, Liu L, Li S, Yang X, Wang S (2022) High-performance lanthanum-based metal–organic framework with ligand tuning of the microstructures for removal of fluoride from water. J Colloid Interface Sci 607:1762–1775
51. Babu K, Rambabu K, Bharath G, Hai A, Hasan SW, Banat F (2023) Biocompatible nanomaterials for sensing and remediation of nitrites and fluorides from polluted water. In: Advances in nano and biochemistry. Elsevier, Amsterdam
52. Solomon-Negrete O, Rodríguez-Hernández AI, Sánchez-Cruz R, Rangel-Mendez JR (2022) Avocado seed-derived biochars for the removal of fluoride from aqueous solutions. Environ Sci Pollut Res 29:38742–38755
53. Manami RB, Megalamani MB, Patil YN, Kalkhambkar RG, Nandibewoor ST, Adarakatti PS, Alsuhaibani AM (2025) Electrochemical investigation of a ZnO/RGO-modified electrode for trace Pb (II) detection in water and soil samples. Top Catal. https://doi.org/10.1007/s11244-025-02087-y
54. Chen Y, Zhao P, Liang Y, Ma Y, Liu Y, Zhao J, Huo D (2023) A sensitive electrochemical sensor based on 3D porous melamine-doped rGO/MXene composite aerogel for the detection of heavy metal ions in the environment. Talanta 256:124294
55. Sivaji SP, Vinothini S, Chen TW, Chen SM, Mariappan K, Chiu TW, Elshikh MS (2025) Newly discovered In2O3@ CNF nanocomposite sensor for highly sensitive electrochemical detection of carbendazim in fruit samples. Microchem J 208:112208
56. Udayan APM (2023) Electrochemical detection of herbicide atrazine using porous MnO2-NiO nanocatalyst. Mater Sci Eng B 290:116302
57. Traiwatcharanon P, Pon-On W, Zacharias M, Wongchoosuk C (2024) Electrochemical copper oxide nanoparticles-based sensor for butachlor plus propanil herbicide detection. J Mater Sci Mater Electron 35(9):651
58. Nonaka LH, Almeida TS, Aquino CB, Domingues SH, Salvatierra V, Souza VH (2020) Crumpled graphene decorated with manganese ferrite nanoparticles for hydrogen peroxide sensing and electrochemical supercapacitors. ACS Appl Nano Mater 3(5):4859–4869
59. Yao L, Yan Y, Lee JM (2017) Synthesis of porous Pd nanostructure and its application in enzyme-free sensor of hydrogen peroxide. ACS Sustain Chem Eng 5(2):1248–1252
60. Ding A, Liu F, Zheng J, Chen J, Li C, Wang B (2018) Synthesis of manganese oxide embedded carbon nanofibers as effective biomimetic enzymes for sensitive detection of superoxide anions released from living cells. Macromol Mater Eng 303(6):1800079
61. Megalamani MB, Patil YN, Chavan C, Nandibewoor ST (2024) A novel ceramic modified carbon-based sensor for ultrasensitive electro-sensing of fast green FCF analysis: an advancement in toxic dye detection. Inorg Chem Commun 163:112288

62. Huang J, Li J, Zhang Z, Li J, Cao X, Tang J, Pan X (2022) Bimetal Ag NP and au NC modified In2O3 for ultra-sensitive detection of ppb-level HCHO. Sens Actuators B Chem 373:132664
63. Panda D, Nandi A, Datta SK, Saha H, Majumdar S (2016) Selective detection of carbon monoxide (CO) gas by reduced graphene oxide (rGO) at room temperature. RSC Adv 6(53):47337–47348
64. Shi H, Fu L, Chen F, Zhao S, Lai G (2022) Preparation of highly sensitive electrochemical sensor for detection of nitrite in drinking water samples. Environ Res 209:112747
65. Balasubramanian P, Annalakshmi M, Chen SM, Chen TW (2019) Ultrasensitive non-enzymatic electrochemical sensing of glucose in noninvasive samples using interconnected nanosheets-like NiMnO3 as a promising electrocatalyst. Sensors Actuators B Chem 299:126974
66. Zalke JB, Narkhede NP, Rotake DR, Singh SG (2024) Facile chemiresistive biosensor functionalized with PANI/GOx and novel green synthesized silver nanoparticles for glucose sensing. Microchem J 200:110339
67. Dai J, Huang JH, Gao LF, Qi ZK, Chen KF (2025) Green synthesis of Nano-ZnO/N-doped porous carbon composites for high-performance non-enzymatic electrochemical glucose sensing. Microchem J 212:113272
68. Patil YN, Megalamani MB, Nandibewoor ST (2024) A novel nanozyme doped ZnO/r-GO-based sensor for highly sensitive electrochemical determination of muscle-relaxant drug: cyclobenzaprine HCl. Microchim Acta 191(6):336
69. Patil YN, Nandibewoor ST (2024) Development of novel Ce doped ZnO/graphene-based sensor for electrochemical investigation of potassium-competitive acid blocker: Vonoprazan. Mater Sci Semicond Process 171:108039
70. Megalamani MB, Patil YN, Nandibewoor ST (2023) YSZ/MoS2 modified carbon-based sensor for the determination of muscle relaxant agent chlorzoxazone: a novel electroanalytical strategy. Inorg Chem Commun 155:111074
71. El-Zahry MR (2024) Electrochemical sensor based on molybdenum-doped graphene oxide nanorods anchored carbon spheres/vanadium pentoxide nanocomposites for simultaneous determination of diclofenac sodium and posaconazole. Microchem J 206:111593
72. Yazdanian M, Sabeti B, Chekin F (2025) Magnetite MXene-based hybrid platforms for electrochemical sensing of anticancer drug Idarubicin. J Electron Mater 54(3):1645–1652
73. Ismardi A, Gunawan LA, Winarsih S, Jatmika J, Manurung RV, Fathona IW, Fariz BN (2024) Detection of highly sensitive Escherichia coli (E. coli) bacteria using an electrochemical sensor based on zinc oxide/polyvinyl alcohol nanocomposite. Microchem J 207:112044
74. Lin X, Liu C, Lei Q, Nan X, Zhu Y, Liao J, Huang Q (2025) A novel ratiometric electrochemical aptasensor based on graphene quantum dots/cu-MOF nanocomposite for the on-site determination of *Staphylococcus aureus*. J Hazard Mater 485:136845
75. Santhosh BM, Rohini BA, Mahanthesha KR, Shivarudrappa HP, Hareesha N, Rashmi CK, Raviraj MK (2025) Recent developments on electrochemical sensors for neurotransmitters using composite transition metal oxide nanoparticles. J Inorg Organomet Polym Mater:1–17. https://doi.org/10.1007/s10904-025-03677-9
76. Harshavardhan A, Santhosh BM, Harshavardhan A, Matt SB, Jambaladinni S, Sidlinganahalli M, Prakash KK (2024) Electrochemical detection of dopamine in the presence of uric acid at cerium oxide and L-tyrosine mixture modified carbon paste electrode: a cyclic voltammetry study. J Electrochem Soc 171:017506. https://doi.org/10.1149/1945-7111/ad1c10

Chapter 13
Summary, Real World Applications, Commercialization and Scalabilty

Yashwanth V. Naik and Ramesh Shankaranarayana

Abstract This chapter provides a comprehensive synthesis of the core advancements, translational potential, commercialization strategies, and scalability challenges related to biocompatible nanomaterial-based smart sensors. It begins by summarizing key material innovations, synthesis approaches, and characterization techniques discussed throughout the book, emphasizing their role in enabling multifunctional, eco-friendly sensor systems. The chapter highlights real-world applications across domains such as heavy metal detection, biomolecular sensing, pesticide monitoring, and pharmaceutical diagnostics, showcasing their relevance in healthcare, agriculture, and environmental safety. Commercialization opportunities are explored through strategies for product development, regulatory compliance, and market entry, alongside the role of academic-industry partnerships and open innovation models. Scalability and sustainable design are addressed through discussions on green manufacturing, circular design principles, and ethical material sourcing. Looking forward, the chapter envisions decentralized, integrated sensing networks aligned with the One Health paradigm, where real-time data from humans, animals, and ecosystems drive informed decision-making. The conclusion underscores the transformative potential of these technologies and calls for multidisciplinary collaboration, responsible innovation, and systemic support to realize a sustainable and sensor-enabled future.

Keywords Sensor · Healthcare · Agriculture

Y. V. Naik (✉)
Department of Physics, RV Institute of Technology and Management, Bangalore, India

R. Shankaranarayana
College of Engineering and Technology, University of Technology and Applied Sciences, Nizwa, Sultanate of Oman

R. Naik et al. (eds.), *Biocompatible Nanomaterials for Smart Sensors*, Nanostructure Science and Technology,
https://doi.org/10.1007/978-3-032-01745-1_13

13.1 Introduction

This concluding chapter consolidates the key advances explored throughout the book, providing a comprehensive overview that bridges foundational research with practical, real-world applications. The journey from laboratory-scale innovations in biocompatible nanomaterials to their integration into advanced smart sensor systems is traced through a multidisciplinary lens. Central to this transformation are the breakthroughs in green and scalable synthesis techniques, precise and high-resolution characterization methods, and the development of multifunctional sensor architectures capable of detecting a wide range of chemical and biological analytes [1]. These innovations collectively underpin a new generation of sensor technologies that are not only highly sensitive and selective but also eco-friendly, cost-effective, and suitable for deployment in diverse settings [2]. The chapter further emphasizes how the convergence of material science with digital technologies such as AI, IoT, and energy-harvesting systems has enabled the creation of autonomous, connected, and intelligent sensing platforms [3]. These systems address critical global challenges, including early disease diagnosis, environmental pollution detection, food safety assurance, and precision agriculture [4]. By reflecting on the translational impact and commercialization potential of these technologies, this chapter aims to provide readers with a roadmap for future research, policy development, and industry collaboration in the field of sustainable smart sensing.

13.2 Summary of Core Advancements

13.2.1 Material Innovations

Chapters 1–4 established the essential scientific and technological groundwork necessary for the development and deployment of biocompatible nanomaterials in smart sensor systems. Chapter 1 introduced the foundational concept of biocompatible nanomaterials, emphasizing their critical role in creating safe, sustainable, and biologically harmonious interfaces for both environmental and biomedical applications [5]. These materials are engineered to avoid toxicity while maintaining high performance, making them ideal for integration into systems that interact directly with living organisms or ecosystems [1]. Chapter 2 expanded upon this foundation by examining the rising significance of multifunctional sensors capable of detecting and responding to a variety of physical, chemical, and biological stimuli within a single platform [6]. The chapter underscored how advances in nanomaterials such as carbon nanotubes, graphene, MXenes, and MOFs combined with AI, IoT connectivity, and energy-harvesting capabilities, are transforming traditional sensors into intelligent, self-sustaining, and highly versatile devices [7]. This multifunctionality not only enhances performance but also reduces device footprint, making sensors more efficient, adaptive, and suitable for real-time applications [8].

Chapter 3 then detailed the innovation in synthesis methods, focusing on environmentally friendly and scalable approaches that align with global sustainability goals [9]. Techniques such as plant-extract-mediated synthesis, microwave-assisted methods, and bio-template fabrication are highlighted for their ability to produce high-quality nanomaterials with minimal ecological impact [10]. These green routes reduce reliance on hazardous chemicals, lower production costs, and open pathways for sustainable industrial-scale manufacturing [11]. Chapter 4 complemented these efforts by presenting a suite of advanced characterization techniques including atomic force microscopy (AFM) [12], Fourier-transform infrared spectroscopy (FTIR) [13], Raman spectroscopy, and X-ray photoelectron spectroscopy (XPS) [14], which are critical for understanding the morphology, composition, and functional properties of nanomaterials. These tools not only validate the structural integrity and performance of synthesized materials but also play a vital role in ensuring consistency, reproducibility, and reliability required for regulatory approval and commercial application [15]. Together, these four chapters created a robust platform for the rational design, sustainable fabrication, and precise evaluation of biocompatible nanomaterials, paving the way for their successful implementation in next-generation smart sensor technologies.

13.3 Translational Applications Across Sensor Domains

Chapters 5–12 comprehensively demonstrated the wide-ranging translational potential of biocompatible nanomaterials, showcasing their applicability across critical domains such as environmental monitoring, healthcare diagnostics, agriculture, and pharmaceutical analysis [16]. In the domain of heavy metal ion detection (Chaps. 5 and 6), sensors based on green-synthesized nanocomposites exhibited ultra-sensitive detection capabilities, reaching parts-per-billion (ppb) levels for hazardous ions like Hg^{2+}, Pb^{2+}, and Cd^{2+} [17]. These innovations are particularly impactful for portable water quality monitoring systems in rural, industrial, and mining-affected regions, where centralized testing infrastructure may be limited. Additionally, the research addressed key challenges such as selectivity in the presence of competing ions, sensor longevity, and the integration of these sensing elements into cost-effective, flexible substrates for real-world deployment [18].

In the field of biomolecular sensing (Chaps. 7 and 8), the focus shifted toward noninvasive, point-of-care diagnostics for clinically relevant metabolites such as uric acid, urea, and creatinine markers vital for kidney function assessment and metabolic disorder management [19]. The incorporation of artificial intelligence (AI) and Internet of Things (IoT) technologies further enhances these systems by enabling real-time data processing, predictive analytics, and cloud-based remote diagnostics [20]. This level of smart integration not only improves diagnostic accuracy and responsiveness but also facilitates scalable telemedicine platforms, particularly beneficial in underserved or remote areas.

Pesticide detection (Chaps. 9 and 10) highlighted the development of eco-friendly nanomaterials and their application in designing compact, rugged sensors capable of identifying common agricultural toxins such as organophosphates and carbamates. These sensors offer enhanced sensitivity, rapid response times, and operational stability, making them highly suitable for on-site, real-time monitoring in agricultural environments [21]. Such innovations support the goals of precision agriculture and food safety, allowing farmers and regulators to respond promptly to contamination events and minimize environmental impact.

Lastly, pharmaceutical drug sensing (Chaps. 11 and 12) explored the use of nanostructured electrochemical sensors for detecting trace levels of antibiotics in both clinical and environmental settings—an essential step in combating the growing threat of antimicrobial resistance [22]. Moreover, biosensors designed for anti-cancer drug monitoring present a powerful tool in therapeutic management, allowing real-time assessment of drug levels during chemotherapy to ensure efficacy while minimizing side effects. These applications underscore the promise of biocompatible sensors in supporting personalized medicine, where treatment strategies are tailored based on dynamic, patient-specific data [23].

Collectively, these chapters highlight how interdisciplinary innovation in materials science, electronics, and data integration is transforming fundamental nanotechnology research into real-world tools that address pressing societal and environmental challenges.

13.4 Commercialization Opportunities and Strategies

Commercializing biocompatible nanomaterial-based sensors presents a transformative opportunity to bring laboratory innovations into practical, market-ready solutions, but it demands a carefully coordinated strategy across the full product development lifecycle. The journey begins with the optimization of scalable and eco-conscious synthesis protocols, such as hydrothermal methods, microwave-assisted techniques, and biotemplating routes that minimize environmental impact while ensuring material quality [24]. Ensuring batch-to-batch reproducibility, chemical stability, and long-term shelf life is crucial for maintaining performance consistency, especially when transitioning from small-scale laboratory production to industrial-scale manufacturing.

Equally important is the design of user-centric interfaces that enable seamless interaction between the sensor and the end-user. This includes developing compact, ergonomic hardware and intuitive software platforms compatible with smartphones, wearable devices, and portable analytical readers. Such integration not only enhances the ease of use and accessibility for a broad spectrum of users from clinicians to farmers and consumers but also facilitates real-time data sharing and cloud-based analytics, aligning with the global push toward digital health and smart environmental management [25].

However, successful commercialization is contingent upon navigating complex regulatory landscapes. Sensors intended for clinical diagnostics or environmental monitoring must comply with stringent international and national standards, including FDA (U.S. Food and Drug Administration) approval for medical devices, CE marking for market access in the European Union, and BIS (Bureau of Indian Standards) certification in India [26]. This regulatory compliance ensures safety, efficacy, and public trust, but also requires the establishment of standardized testing frameworks and certification pipelines, particularly for emerging materials like green-synthesized nanocomposites, which lack universally accepted benchmarks [27].

The market potential for these sensors is expansive and rapidly growing, with major demand across healthcare diagnostics (e.g., point-of-care biosensors for chronic disease monitoring), precision agriculture (e.g., real-time soil and pesticide sensors), and environmental monitoring (e.g., detection of pollutants in air and water). Each of these sectors benefits from the convergence of sustainability, miniaturization, and digital connectivity, underscoring the relevance of biocompatible and multifunctional sensor platforms [2].

To effectively bridge the gap between research and real-world deployment, open innovation models are critical. These include academic spin-offs, which translate university-based discoveries into commercial ventures; public-private partnerships, where industries co-develop technologies with research institutions; and government-funded accelerators and grant programs that provide essential seed funding, infrastructure, and policy support [28]. Such collaborative ecosystems are vital for overcoming financial, infrastructural, and regulatory hurdles, thereby accelerating the deployment of next-generation sensor technologies that are not only functional and cost-effective but also ethically and environmentally sustainable.

13.5 Scalability Challenges and Sustainable Design

The scalability of biocompatible nanomaterial-based sensors is a pivotal concern as innovations transition from research laboratories to commercial and industrial applications. This transition is often impeded by a range of manufacturing and sustainability challenges that must be addressed through deliberate and forward-thinking strategies [29]. One of the foremost issues lies in the high cost and limited availability of raw biomaterials, many of which are derived from rare or sensitive biological sources. Additionally, scaling up green synthesis techniques, which are often optimized for small-scale or batch processes, presents substantial difficulties in maintaining reaction control, uniformity, and purity at higher production volumes. Ensuring nanoparticle stability during large-scale fabrication and storage is another technical hurdle, as fluctuations in size, morphology, or surface chemistry can compromise sensor performance and reliability [30].

To address these challenges, innovative material sourcing strategies are being explored, particularly the utilization of low-cost, renewable biomass such as agro-waste (e.g., rice husk, banana peel) and marine biopolymers like chitosan and cellulose [31]. These materials offer not only economic advantages but also environmental benefits, aligning with the goals of sustainable development. Furthermore, the adoption of modular sensor designs that support plug-and-play manufacturing frameworks can significantly streamline production, reduce assembly complexity, and improve maintenance and scalability. Such modularity also facilitates customization for diverse applications without overhauling the entire manufacturing process.

Beyond production, there is a critical need to consider the entire lifecycle of the sensor, encompassing raw material extraction, fabrication, operational use, and eventual disposal. This calls for an integrated environmental and ethical framework that supports eco-design and circular economy principles. Key strategies include using biodegradable substrates to minimize plastic waste, developing recyclable or reusable electrode materials, and promoting waste-free synthesis processes that reduce chemical residues and emissions. These practices not only reduce the ecological footprint of sensor technologies but also support regulatory compliance and public acceptance, particularly in sectors where environmental impact is a major concern [32].

Ultimately, embedding sustainability into the core of sensor development is not merely a regulatory or moral imperative it is a strategic enabler of long-term commercial viability. As markets and consumers become more environmentally conscious, sensors designed with ethical sourcing, green manufacturing, and end-of-life responsibility will stand out as next-generation solutions capable of meeting the dual demands of technological sophistication and ecological stewardship.

13.6 Future Outlook: Toward Decentralized and Integrated Sensing Networks

The future trajectory of biocompatible nanomaterial-based sensors is increasingly aligned with the development of decentralized and integrated sensing networks, poised to revolutionize how we monitor, understand, and respond to health and environmental conditions. These next-generation systems are designed to be autonomous, intelligent, and highly responsive, enabling continuous, real-time monitoring without the constraints of centralized infrastructure [33]. A significant leap forward in this vision is their seamless integration with smart systems, wherein sensors are embedded directly into everyday materials such as wearable textiles, smart packaging, building materials, and biomedical implants. This convergence facilitates unobtrusive data collection in real-time, unlocking transformative applications such as wearable health trackers that provide continuous vitals monitoring, intelligent food packaging that indicates spoilage, and implantable diagnostics that autonomously track biochemical changes within the human body.

The rise of citizen science and community-based diagnostics further amplifies the societal impact of these sensors. By offering low-cost, portable, and easy-to-use sensor kits derived from biocompatible nanomaterials, these tools empower individuals and underserved communities to actively participate in monitoring local health risks, environmental pollution, and water quality [34]. This democratization of sensing technology ensures that timely data is no longer limited to centralized laboratories or institutional hubs, but is instead available at the grassroots level, enabling faster, locally-driven decision-making and improved preparedness for public health emergencies.

Moreover, the increasing adoption of the One Health paradigm which emphasizes the interconnectedness of human, animal, and environmental health positions biocompatible sensors as pivotal instruments for ecosystem-wide diagnostics and surveillance. By capturing and correlating data across species and natural systems, these sensors can detect zoonotic disease outbreaks, monitor biodiversity loss, and track environmental stressors such as climate-induced pollution or habitat degradation. Such holistic data collection is critical for shaping evidence-based policy, deploying early warning systems, and fostering sustainable coexistence between human societies and the natural world [35].

To support this complex, data-driven ecosystem, these sensor networks are increasingly being integrated with advanced computational tools such as machine learning, edge computing, and blockchain-based data security, ensuring not only accurate interpretation but also trust, privacy, and scalability in data sharing. As technology continues to evolve, biocompatible sensors will be central to building resilient, adaptive, and health-centric infrastructures that can respond to emerging global challenges [36]. Their interdisciplinary nature, drawing from materials science, biology, engineering, and data analytics, underscores their versatility and promise in contributing to a more inclusive, informed, and sustainable future.

13.7 Conclusion

Biocompatible nanomaterials have emerged as a transformative force in the development of a new generation of smart sensors devices that are not only highly sensitive and selective, but also sustainable, safe, and compatible with biological and ecological systems. These materials, owing to their nontoxic nature, green synthesis routes, and versatile functional properties, have enabled the creation of multifunctional sensing platforms that can operate in complex environments, whether inside the human body, in agricultural fields, or across polluted ecosystems. As this book has demonstrated through an exploration of cutting-edge research, practical case studies, and translational applications across domains, the journey from fundamental science to real-world implementation is both challenging and deeply rewarding.

Achieving widespread adoption and impact, however, depends on more than just scientific and technological breakthroughs. It requires cross-disciplinary collaboration among material scientists, engineers, clinicians, environmental scientists, and policymakers. Moreover, regulatory alignment, standardized testing protocols, and strong industry-academic partnerships are vital to move these technologies from the lab bench to the marketplace. Ethical considerations and sustainability principles must guide every stage of the innovation lifecycle from sourcing raw materials and designing sensor architectures to deployment, data management, and end-of-life disposal.

The potential of biocompatible nanomaterial-based sensors is nothing short of revolutionary. They promise a world where real-time, autonomous sensing is embedded in the very fabric of our daily lives within wearables that monitor health continuously, water systems that self-report contamination, crops that signal nutrient deficiencies, and ecosystems that communicate signs of stress. This seamless integration of sensing technologies into both natural and built environments heralds a future where health, safety, and sustainability are not reactive concerns but proactive design elements.

Ultimately, this vision demands responsible innovation, supported by inclusive governance and robust public engagement. As we stand at the intersection of technological possibility and societal need, biocompatible sensors offer a pathway toward a smarter, more resilient, and equitable world one where scientific progress aligns with ecological harmony and human well-being.

References

1. Kirubakaran D, Wahid JBA, Karmegam N, Jeevika R, Sellapillai L, Rajkumar M, SenthilKumar KJ (2025) A comprehensive review on the green synthesis of nanoparticles: advancements in biomedical and environmental applications. Biomed Mater Devices. https://doi.org/10.1007/s44174-025-00295-4
2. Chaudhary V, Gaur P, Rustagi S (2024) Sensors, society, and sustainability. Sustain Mater Technol 40:e00952. https://doi.org/10.1016/j.susmat.2024.e00952
3. Simionescu M, Strielkowski W (2024) The role of the internet of things in enhancing sustainable urban energy systems: a review of lessons learned from the COVID-19 pandemic. J Urban Technol. https://doi.org/10.1080/10630732.2024.2411932
4. Nath S (2024) Advancements in food quality monitoring: integrating biosensors for precision detection. Sustain Food Technol. https://doi.org/10.1039/D4FB00094C
5. Ighodaro A, Osarobo JA, Onuguh IC, Ogbeide OK, Ifijen IH (2024) Challenges and future perspectives of biomimetic materials for biomedical applications: bridging the gap between nature and medicine. In: TMS annual meeting & exhibition. Springer, Cham, pp 877–896. https://doi.org/10.1007/978-3-031-50349-8_76
6. Lee HK, Yang YJ, Koirala GR, Oh S, Kim TI (2024) From lab to wearables: innovations in multifunctional hydrogel chemistry for next-generation bioelectronic devices. Biomaterials 310:122632. https://doi.org/10.1016/j.biomaterials.2024.122632
7. Ray PP (2024) A review on IoT energy storage with nanocarbon materials: requirements, state-of-the-art, challenges, and future scope. In: NanoCarbon: a wonder material for energy

applications: volume 2: fundamentals and advancement for energy storage applications. Springer, Cham, pp 41–69. https://doi.org/10.1007/978-981-99-9931-6_3
8. Ma CB, Shang X, Sun M, Bo X, Bai J, Du Y, Zhou M (2025) Emerging multifunctional wearable sensors: integrating multimodal sweat analysis and advanced material technologies for next-generation health monitoring. ACS Sens. https://doi.org/10.1021/acssensors.4c03396
9. Ahmad S, Jaiswal R, Yadav R, Verma S (2024) Recent advances in green chemistry approaches for pharmaceutical synthesis. Sustain Chem 4:100029. https://doi.org/10.1016/j.scowo.2024.100029
10. Pareek P, Saini H, Ledwani L (2025) Green synthesis of tin oxide (SnO_2) nanoparticles using pomelo fruit peel for photocatalytic degradation of methylene blue dye and industrial dye. Top Catal 68(3):252–272. https://doi.org/10.1007/s11244-024-02031-6
11. Papanikolaou G, Centi G, Perathoner S, Lanzafame P (2024) Green synthesis and sustainable processing routes. Curr Opin Green Sustain Chem 47:100918. https://doi.org/10.1016/j.cogsc.2024.100918
12. She DT, Nai MH, Lim CT (2024) Atomic force microscopy in the characterization and clinical evaluation of neurological disorders: current and emerging technologies. Med-X 2(1):8. https://doi.org/10.1007/s44258-024-00022-6
13. Valente PA, Mota SI, Teixeira A, Ferreiro E, Sarmento H, Cipriano I, Oliveira PJ (2025) Fourier transform infrared (FTIR) spectroscopy as a tool to characterize exercise and physical activity: a systematic review. Sports Med 55(2):459–472. https://doi.org/10.1007/s40279-024-02139-5
14. Anekwe IMS, Akpasi SO, Enemuo EM, Ashiegbu D, Mustapha SI, Isa YM (2024) Innovations in catalytic understanding: a journey through advanced characterization. Mater Today Catal 7:100061. https://doi.org/10.1016/j.mtcata.2024.100061
15. Singh AB, Khandelwal C, Dangayach GS (2024) Advancements in healthcare materials: unraveling the impact of processing techniques on biocompatibility and performance. Polym Plast Technol Mater 63(12):1608–1644. https://doi.org/10.1080/25740881.2024.2350026
16. Fakayode SO, Lisse C, Medawala W, Brady PN, Bwambok DK, Anum D, Grant C (2024) Fluorescent chemical sensors: applications in analytical, environmental, forensic, pharmaceutical, biological, and biomedical sample measurement, and clinical diagnosis. Appl Spectrosc Rev 59(1):1–89. https://doi.org/10.1080/05704928.2023.2177666
17. Sharma V, Dutta S, Roy RK, Manna S, Choudhury SM, Patra GK (2025) Highly sensitive benzildihydrazone-N, N'-bis (2-hydroxy-4-diethylamino-1-formylbenzene) stabilized ZnS nanoparticles as potential optical chemosensors for Hg^{2+} ions: anticancer activity and biosensor imaging. J Mol Struct 1329:141398. https://doi.org/10.1016/j.molstruc.2025.141398
18. Leburu E, Qiao Y, Wang Y, Yang J, Liang S, Yu W, Hou H (2024) Flexible electronics for heavy metal ion detection in water: a comprehensive review. Biomed Microdevices 26(3):30. https://doi.org/10.1007/s10544-024-00710-5
19. Çelik H, Caf BB, Geyik C, Çebi G, Tayfun M (2024) Enhancing urinalysis with smartphone and AI: a comprehensive review of point-of-care urinalysis and nutritional advice. Chem Pap 78(2):651–664. https://doi.org/10.1007/s11696-023-03137-z
20. Aminizadeh S, Heidari A, Dehghan M, Toumaj S, Rezaei M, Navimipour NJ, Unal M (2024) Opportunities and challenges of artificial intelligence and distributed systems to improve the quality of healthcare service. Artif Intell Med 149:102779. https://doi.org/10.1016/j.artmed.2024.102779
21. Xie A, Zhou Q, Fu L, Zhan L, Wu W (2024) From lab to field: advancements and applications of on-the-go soil sensors for real-time monitoring. Eurasian Soil Sci 57(10):1730–1745. https://doi.org/10.1134/S1064229324601124
22. Sajwan RK, Hashmi SZH, Himanshu JK, Kumari A, Solanki PR (2024) Current advancement in nanomaterial-based emerging techniques for the determination of aminoglycosides antibiotics for antibiotic resistance surveillances. Mater Adv 5(3):961–985. https://doi.org/10.1039/D3MA00632H

23. Mazumdar H, Khondakar KR, Das S, Halder A, Kaushik A (2025) Artificial intelligence for personalized nanomedicine; from material selection to patient outcomes. Expert Opin Drug Deliv 22(1):85–108. https://doi.org/10.1080/17425247.2024.2440618
24. Jabbar Khan A, Gao L, Numan A, Khan S, Hussain I, Sajjad M, Zhao G (2024) Recent advancements in the tailoring of nanomaterials via microwave-assisted synthesis: a comprehensive review. Crit Rev Solid State Mater Sci. https://doi.org/10.1080/10408436.2024.2446049
25. Kumari A, Khatoon Z, Kumrai S, Jain A (2024) Sustainable development through real-time monitoring of healthcare data analytics on the cloud. In: International conference on sustainable development through machine learning, AI and IoT. Springer, Cham, pp 254–262. https://doi.org/10.1007/978-3-031-71729-1_23
26. Chattopadhyay P, Goyary D (2024) Modern medicine: biomedical devices, medical gases, radiopharmaceuticals, new drug discovery, vol 2. Routledge, London. https://doi.org/10.1201/9781003397854
27. Arif MD, Hoque ME, Rahman MZ, Shafoyat MU (2024) Emerging directions in green nanomaterials: synthesis, physicochemical properties and applications. Mater Today Commun 40:109335. https://doi.org/10.1016/j.mtcomm.2024.109335
28. Owen R, Vedanthachari LN, Hussain J (2024) The role of the university entrepreneurial ecosystem in entrepreneurial finance: case studies of UK innovation knowledge centres. Ventur Cap 26(3):351–375. https://doi.org/10.1080/13691066.2023.2205606
29. Elgin C (2025) Anticipatory macroeconomic governance: exploring future-oriented strategies for economic resilience and sustainability. Eur J Futures Res 13(1):4. https://doi.org/10.1186/s40309-025-00248-6
30. Thakur A, Kumar A (2024) Challenges and opportunities in the development of nano-hybrid smart coatings. In: Nano-hybrid smart coatings: advancements in industrial efficiency and corrosion resistance. ACS Publications, Washington, pp 353–384. https://doi.org/10.1021/bk-2024-1469.ch015
31. Oyedeji S, Patel N, Krishnamurthy R, Fatoba PO (2024) Agricultural wastes to value-added products: economic and environmental perspectives for waste conversion. Springer, Cham. https://doi.org/10.1007/10_2024_274
32. Dai S (2025) Understanding automation's impact on ecological footprint: theory and empirical evidence from Europe. Environ Resour Econ 88(2):503–532. https://doi.org/10.1007/s10640-024-00938-y
33. Gupta S, Verma R, Dhanda N (2024) Introduction to next-generation internet and distributed systems. Decentral Syst Distrib Comput. https://doi.org/10.1002/9781394205127.ch1
34. Srinivas A, Chandran KR, Nehra S, Kumar D (2025) Future aspects in the field of bio-nanomaterials toward environmental assessment. In: Bio-nanomaterials in environmental remediation: industrial applications. Wiley, Hoboken, pp 295–317. https://doi.org/10.1002/9783527848546.ch12
35. Chakravarty P, Gattupalli S, Chakravarty U, Chand G, Lee W (2025) Advancing earth resource management through AI enhanced early warning systems and crisis communication. In: Data analytics and artificial intelligence for earth resource management. Elsevier, Amsterdam, pp 77–95. https://doi.org/10.1016/B978-0-443-23595-5.00005-X
36. Gideon O, Samuel HS, Okino IA (2024) Biocompatible materials for next-generation biosensors. Discov Chem 1(1):34. https://doi.org/10.1007/s44371-024-00042-6

Index

R. Naik et al. (eds.), *Biocompatible Nanomaterials for Smart Sensors*, Nanostructure Science and Technology, https://doi.org/10.1007/978-3-032-01745-1

FSC
www.fsc.org
MIX
Papier aus verantwortungsvollen Quellen
Paper from responsible sources
FSC® C105338

If you have any concerns about our products,
you can contact us on
ProductSafety@springernature.com

In case Publisher is established outside the EU,
the EU authorized representative is:
Springer Nature Customer Service Center GmbH
Europaplatz 3, 69115 Heidelberg, Germany

Printed by Libri Plureos GmbH
in Hamburg, Germany